SpringerBriefs in Philosophy

Philosophy of Science

Series editor

Sahotra Sarkar, Austin, TX, USA

Editorial Advisory Board

Hannes Leitgeb, Ludwig-Maximilians-Universität, München, Germany
Samir Okasha, University of Bristol, Bristol, UK
Laura Ruetsche, University of Michigan, Ann Arbor, USA
Andrea Woody, University of Washington, Seattle, USA

Managing Editor

Derek Anderson

SpringerBriefs in Philosophy of Science will consist of original works in the philosophy of science that present new research results or techniques that are of broad instructional significance at the graduate and advanced undergraduate levels.

Topics covered will include (but not be limited to):

Formal epistemology
Cognitive foundations of science
Metaphysics of models, laws, and theories
Philosophy of biology
Philosophy of chemistry
Philosophy of mathematics
Philosophy of physics
Philosophy of psychology
Philosophy of the social sciences

The series is intended to bridge the gap between journal articles and books and monographs. Manuscripts that are too long for journals but either too specialized or too short for books will find their natural home in this series. These will include suitably edited versions of lectures and workshop presentations that develop original perspectives on the philosophy of science that merit wide circulation because of the novelty of approach.

The length of each volume will be 75–125 published pages.

More information about this series at http://www.springer.com/series/13349

Justin Garson

A Critical Overview
of Biological Functions

 Springer

Justin Garson
Department of Philosophy
Hunter College of the City University
  of New York
New York, NY
USA

ISSN 2211-4548   ISSN 2211-4556   (electronic)
SpringerBriefs in Philosophy
ISSN 2366-4495   ISSN 2366-4509   (electronic)
Philosophy of Science
ISBN 978-3-319-32018-2   ISBN 978-3-319-32020-5   (eBook)
DOI 10.1007/978-3-319-32020-5

Library of Congress Control Number: 2016935570

Printed on acid-free paper

This Springer imprint is published by Springer Nature
The registered company is Springer International Publishing AG Switzerland

Acknowledgements

I am grateful to several people who provided comments on parts of this manuscript: Paul Sheldon Davies, Matteo Mossio, Bence Nanay, and Gerhard Schlosser. I would like to give special thanks to Derek Anderson, Dan McShea, Gualtiero Piccinini, and an anonymous referee for providing extensive comments on the entire manuscript. For several discussions about function that have informed the ideas in this volume, I would like to thank Carl Craver, Dan McShea, Karen Neander, Gualtiero Piccinini, and Sahotra Sarkar. I am also grateful to Sahotra Sarkar for suggesting that I write this volume for the SpringerBriefs in Philosophy of Science series.

Acknowledgements

Contents

Chapter 1
What Is a Theory of Function Supposed to Do?

Abstract Biological functions are central to several debates in science and philosophy. In science, they play a role in debates in genetics, neuroscience, biomedicine, and ecology. In philosophy, they play a role in debates about the nature of teleological reasoning, biological information, trait classification, normativity, meaning and mental representation, health, disease, and the nature of artifacts. Yet philosophers and scientists disagree about what biological functions are, or whether there are different kinds of functions. One problem is that they do not agree about what, precisely, a philosophical theory of biological function is supposed to be or to do. I begin this chapter by discussing why functions matter to philosophy and science. I lay out three very traditional desiderata for a theory of function: namely, that the theory account for the function/accident distinction and the explanatory and normative features of function. I review three main approaches to thinking about what a theory of function should be: a conceptual analysis, a theoretical definition, and a Carnapian-style explication. I argue that it does not matter which one we accept so long as we agree that a theory of function should be reasonably constrained by actual biological usage.

Keywords Biological functions · Conceptual analysis · Theoretical definition · Normativity of function · Artifact functions

1.1 Why Functions Matter to Biology and Philosophy

Functions matter to biology. For example, geneticists recently formed a large consortium called ENCODE (ENCyclopedia Of DNA Elements). The scope and ambitions of this project rival that of the Human Genome Project. Their stated mission is to sequence the functional elements of the human genome. One of the claims ENCODE's proponents repeatedly make is that over 80 % of the human genome is functional (ENCODE Project Consortium 2012). This is supposed to debunk the popular claim that only 10 % of the human genome is functional and the rest constitutes "junk DNA."

© The Author(s) 2016   1
J. Garson, *A Critical Overview of Biological Functions*,
Philosophy of Science, DOI 10.1007/978-3-319-32020-5_1

Scientific critics of ENCODE, however, claim that this 80 % estimate is a mistake. It rests, they charge, on an extremely liberal construal of the term "function" that contradicts standard biological usage (Graur et al. 2013; Doolittle 2013). As the geneticist Ford Doolittle puts it, it involves an "unacknowledged conflation of possible meanings of 'function'" (2013, 5294). The scientists leading the consortium recently responded to these attacks by arguing that there are multiple meanings of "function" in biology, all of which are legitimate (Kellis et al. 2014). A recent editorial in the journal *Nature* admits that ENCODE raises deep questions about the very *concept* of function.[1] The problem, according to the editorialist, is that biologists do not spend very much time trying to define "function" in any explicit way.

Functions also matter to medicine, psychiatry and psychiatric classification. Over the last 40 years, psychiatrists have repeatedly engaged in debates about psychiatric classification, that is, about whether or not such-and-such constitutes a legitimate mental disorder. Recent debates include whether or not transgender people have "mental disorders," whether pre-menstrual syndrome (PMS) can ever be a "mental disorder," or whether the experience of grief (for example, following the loss of a loved one) can constitute a "mental disorder" (see American Psychiatric Association 2013). These debates are often divisive for psychiatric professionals as well as society at large.

Some scientists have argued that we should define "mental disorder" explicitly in terms of biological function. The idea is that what makes something a mental disorder is that it is caused by an inner dysfunction on the part of the individual. Such a definition, they argue, would have valuable implications for psychiatric classification and research (Cosmides and Tooby 1999; First 2007; Nesse 2007). These psychiatrists have directly borrowed their ideas about function from the philosophical debates (e.g., Wakefield 1992; Boorse 1977).

The notion of function also matters to ecology and conservation biology. For example, one influential criticism of ecological restoration holds that it is impossible to ever truly restore an ecosystem (e.g., Elliot 1982; Katz 1992). The reason is not just the obvious one, that, due to technical limitations we can never make a current ecosystem just like it was in the past. The problem is a deeper, conceptual one, and it stems from the very concepts of nature and artifact. In theory, whenever human beings redesign an ecosystem, they make it into a kind of artifact of their own plans and purposes, which is the exact opposite of nature. So, it is impossible to restore nature, because the outcome of such an intervention is an unnatural artifact. This controversial argument, however, rests explicitly on a specific (and I think dubious) account of biological functions (Katz 1992, 235; also see Lo 1999; Siipi 2003; Garson Submitted for publication a for discussion). So, fundamental questions about the nature and value of ecological restoration are tied to questions about the meaning of function.

---

[1]Form and function [Editorial] *Nature* 495: 141–142 (March 14, 2013).

Functions also matter to philosophy. They matter to the philosophy of science and biology, the philosophy of medicine, and the philosophy of mind. First, many philosophers of biology have argued that core concepts of biology, such as the concept of biological information, the notion of a biological trait, and even the concept of a biological mechanism, should be defined in terms of function (for discussion of information, see Maynard Smith 2000; Sterelny 2000; Sarkar 2000, 2013; Godfrey-Smith 2007; Stegmann 2009; Shea 2013; for trait, see Neander 2002; Griffiths 2006; Rosenberg and Neander 2009; Nanay 2010; Neander and Rosenberg 2012; for mechanism see Craver 2001, 2013; Nervi 2010; Moghaddam-Taaheri 2011; Moss 2012; Garson 2013; Mebius 2014). The notion of function is one of the foundational concepts of the life sciences. As a consequence, the notion of function touches upon almost every major debate in the philosophy of biology.

Functions matter to the philosophy of medicine. In the 1970s, Christopher Boorse argued that the concept of *disease* should be explicated in terms of biological function (e.g., Boorse 1976, 1977), and his work generated a lively debate that is still very active today (for very recent entries, see Schwartz 2007; Kingma 2010; Hausman 2011; Garson and Piccinini 2014; Boorse 2014). The philosopher Jerome Wakefield (1992) extended Boorse's discussion by arguing that the notion of *mental disorder* should be defined in terms of dysfunction, which also generated an ongoing controversy (see Adriens and de Block 2011; Faucher and Forest Submitted for publication for recent anthologies). These debates have implications for health care policy and biomedical ethics (e.g., Daniels 1985).

Functions matter for the philosophy of mind. One perennial question is the nature of representational content or intentionality: what is it for a mental state to represent a state of affairs? One viewpoint is that we can make sense of representational content in terms of biological function, though controversies remain about how, precisely, to carry out this project (for recent entries here, see Macdonald and Papineau 2006; Neander 2012; Millikan 2004; Ryder Submitted for publication; see Garson 2015, Chap. 7, for an overview). According to one view, the content of a representation should be defined in terms of the biological function of the mechanism that produces the representation ("producer" or "informational" teleosemantics). According to another view, the content of a representation should be defined in terms of the biological function of the mechanisms that utilize or consume the representation ("consumer" teleosemantics). Others maintain that the very attempt to explain mental representation in terms of biological function is hopelessly flawed (Crane 2003; Burge 2010). The functions debate in philosophy is not going away any time soon.

1.2   Goals for a Theory of Function

What are some goals—some desiderata—for a theory of function? What exactly is a theory of function supposed to do? A problem here is that philosophers disagree about what these desiderata are, and that leads to disagreement about which theory of function is correct. The process of devising and articulating desiderata (or, more strictly, adequacy conditions) is a tricky business. That is because they are never entirely neutral. In other words, one's intuitions about whether such-and-such constitutes an appropriate desideratum for a theory of function is partly shaped by one's prior commitments about which theory is right.

The best I can do, given the fact that desiderata are never entirely neutral, is to set out three very conventional desiderata that have appeared in the literature over the last several decades. They have become more-or-less canonical in the philosophical functions debate. We should, however, bear in mind that any of these desiderata, taken individually, are defeasible. In other words, it may be possible to give good reasons for why one of them should not actually constrain our theorizing about function. So, they are not strict adequacy conditions. If someone wishes to reject one of these desiderata, however, that person should provide good reasons for doing so. The same holds if one wants to add a new desideratum to the list.

Most writers on function have assumed that a theory of function should make sense of three features of functions: the distinction between function and accident, the explanatory dimension of functions, and the normative dimension of functions. First, a theory of function should account for the distinction between function and (lucky) accident. The function of my nose is to help me to breathe, but not to hold up my glasses, despite the fact that it does both and both are good for me. The latter is a lucky accident.

Very few theorists have ignored this distinction (though they may contest that particular example). A notable exception is Bock and von Wahlert (1965), who say that the functions of a trait include, "all physical and chemical properties arising from its form" (274). That is the only paper, to my knowledge, that does not even purport to make such a distinction. There are some theorists who accept the distinction but think it is relative to our explanatory goals (see Chap. 5). To relativize the distinction to our explanatory goals is not the same thing as ignoring it altogether.

Second, a theory of function should make sense of the explanatory dimension of function statements. At least sometimes, when we attribute a function to a trait, we purport to explain why the trait is there, that is, why organisms possess the trait. For example, in some contexts, when one says, "the function of eyespots on butterfly wings is to deflect attack away from vital organs," one purports to explain why some butterflies have eyespots. When one says, "the function of zebra stripes is to deter biting flies," one purports to explain why zebras have stripes, rather than, say, being monocolored like horses. Functions do not always purport to be explanatory, but sometimes they do. A good theory of function should make sense of this puzzling feature of biological usage.

There is disagreement, however, about what "explanation" means, and about what exactly functions are supposed to explain (the *explanandum*). One natural way to think about explanation is in terms of causal explanation. In other words, one might think that, to the extent that function statements purport to be explanatory, they purport to be causal explanations for the existence of traits (see Chap. 3, Sect. 4.3, and Sects. 6.1 and 6.2 for discussion). Such explanations are often called "etiological" explanations. An etiological explanation for an event is simply one that explains the event in terms of some prior cause. Since I prefer not to multiply technical terms, I will generally stick with the expression "causal explanation."

I do not claim, however, that *all* explanations must be causal explanations. I think there are other ways of satisfying this explanatory aspect of functions without understanding explanation in the causal sense. Ernest Nagel (1961), for example, believed that function ascriptions are explanations for the existence of traits, but he thought they were explanations in a different, epistemic sense of the term. Specifically, he thought that explaining the existence of a trait amounted to correctly deducing its existence, given some other salient biological facts and general principles. Others have followed Nagel in interpreting this demand for explanation broadly, as I will describe in Sect. 4.3.

Some writers maintain that functions are explanatory, but they do not agree that functions explain the presence of the functionally characterized item. Rather, these writers switch the *explanandum* of the function statement. For example, Walsh (1996) and Walsh and Ariew (1996) believe that functions are explanatory. However, when they say that a trait *currently* has a function (e.g., the zebra's stripes currently have the function of deterring biting flies) they merely purport to explain why we should expect this trait to persist well into the future. They are not explaining why stripes exist today. (We can, however, explain why stripes exist today by citing some function they performed in the past. I will return to their claim in Sect. 4.3.) Cummins (1975), similarly, claims that function statements do not explain the functional item. Rather, they explain the functional effect. For example, when we say that the function of the heart is to circulate blood, we are not explaining why people have hearts. We are explaining how blood circulation works.

Finally, a theory of function should account for the "normativity" of functions, in a special sense of that term. By "normativity" I simply mean that it is logically possible for a trait token to have a function that it cannot, in fact, perform. Henry's stomach has the function of digesting food even though, owing to a severe ulcer, his stomach cannot do so at present (it malfunctions or fails to function). I realize that some people do not like using the term "normativity" in this context because the term makes them think of values and ethics. I use the term in a way that has nothing to do (at least on the surface) with values or ethics. It may turn out that the best way to make sense of the normativity of function is by appeal to values and ethics, but I think we can understand this idea in a way that is independent of it.

Some theorists have contested the validity of this demand for normativity. Some have gone so far as to claim that the notion of malfunctioning is conceptually incoherent (Davies 2000, 2001, 2009). That would be surprising to me given how commonly the notion of malfunctioning is used in biology and biomedicine, but I will discuss this line of thought in Sect. 3.3. Others have argued that the notion of malfunctioning may have troubling political or social implications, particular for people with disabilities (e.g., Amundson 2000). I think this is a legitimate concern (as I indicated in Garson 2015, Chap. 8). I too am wary of the way that psychiatrists throw around the term "dysfunction" because I fear that it can stigmatize certain behaviors or choices. On the other hand, sometimes the notion of malfunctioning is helpful for science, for example, when it works as a heuristic to help biologists uncover the roots of diseases like HIV/AIDS or various diseases of neurulation (Garson 2013). I do not think we should toss the notion of malfunctioning out the window. Instead, we should use it carefully and thoughtfully. I hope that this volume will help us to better reflect on how we might avoid negative social implications of function and dysfunction.

Various writers have suggested other desiderata for a theory of function. I do not believe these recur frequently enough in the literature to be taken as given. The first is that functions not be "epiphenomenal," in a special sense of that term (Mossio et al. 2009, 821). The demand here, if I understand it correctly, is that any good theory of function should have the following implication: when we attribute a function to a trait, we are saying something about its current-day causal powers. One way to support this claim is to say that, if functions are not identical to current-day causal powers of a trait, then they are scientifically useless and irrelevant, because science only cares about understanding the causal powers of entities. Note that, if we accept this desideratum, we should *reject* the selected effects theory of function (and its ilk) because the selected effects theory defines the function of a trait in terms of its history and not its present-day causal powers. (Actually, there are a lot of respectable scientific concepts that are defined in terms of history, like the idea of an adaptation, a volcanic mountain, or a sibling. I do not think science should get rid of those sorts of concepts.)

One concern I have with this desideratum is that there seems to be a tension between it and the normativity of functions. To say that functions are normative is to say that it is possible for a trait token to possess a function that it cannot perform. On the surface, to demand that functions not be epiphenomenal is to demand that, if a trait token has a function, it must be able to perform that function. So it looks like we may have to dismiss one of these two desiderata. It seems to me that the normativity of functions is more deeply entrenched in the functions literature, so if I had to dismiss one, I would dismiss the claim that functions should not be epiphenomenal. Perhaps all that the objection amounts to is that, if a trait has a function, it must be the case that *most* tokens of the trait can perform that function. But this would preclude, by fiat, the possibility of pandemic dysfunction (see Sect. 4.3).

One referee of this volume made the following interesting observation about the epiphenomenal objection. In this referee's opinion, the charge that selected effects functions are epiphenomenal is really part of a deeper, more foundational, worry about selected effects functions. The deeper worry is that the selected effects theory contradicts the way that biologists actually use the term "function." According to this line of thought, when biologists attribute functions to traits, they do not make any reference to the etiology of the trait. They are just describing some features of its current-day behavior.

I deal with this particular objection in a few different places in this volume, but I will provide a very brief response here. There are three points I wish to make. First, I accept that biologists sometimes use the term "function" with no obvious etiological import. But sometimes they do. That is, sometimes they do use "function" with etiological import, and when they do, we should take it seriously. The fact—if it is a fact—that biologists often use the term "function" with no etiological import would lead, at best, to pluralism about biological functions (see Sect. 5.3 for more on pluralism). Second, even when biologists do not *explicitly* refer to a trait's etiology when attributing a function to it, perhaps they often do so *implicitly*. The purpose of a philosophical analysis of function is to explicate the sorts of ontological commitments that biologists are implicitly committed to. I believe when biologists attribute functions to traits, they are often implicitly, if not explicitly, citing selection processes (see Sect. 3.3. under the heading *"Does the Selected Effects Theory Capture the Way Biologists Use the Term?"*). Third, I agree that, sometimes, when biologists are trying to resolve controversies about the function of a trait, they merely appeal to evidence about what the trait *currently* does in a population. For example, when biologists argue with one another about the function of eyespots on butterfly wings, they often examine the way that eyespots currently benefit butterflies. This may lead one to think that functions must be *constituted* by current-day causal powers. But I do not accept this conclusion. Just because biologists appeal to current-day causal powers of a trait in order to resolve controversies about function, that does not mean that functions are constituted by current-day causal powers. It may simply mean that examining a trait's current-day causal powers provides strong, albeit defeasible, evidence about a trait's function (Sect. 5.3).

Two other potential desiderata have made an appearance in the literature. One is that functions should be iterative (for lack of a better term). This means that, if the function of some item I is F, and some component C of I contributes to F, then C should have a function, too (namely, whatever it does that helps I do F). In other words, functions should "trickle down" from the system as a whole to the components of that system (see Kitcher 1993; Buller 1998). Another desideratum, one that was more popular in the 1970s, is the idea that a theory of function should provide a unified treatment of both biological and artifact functions (Wright 1973; Boorse 1976; also see Lewens 2004 for discussion). Many writers, particularly since the 1990s, have abandoned this latter desideratum because they cannot see how a single theory can make sense of both sorts of functions (Neander 1991; Godfrey-Smith 1993). I will come back to this point in Sect. 1.4. This is a good example of how desiderata for a theory of function sometimes change over time.

1.3 Meta-Analytic Aspects: Conceptual Analysis, Theoretical Definition, or What?

In addition to setting out our desiderata, we should also think about what kind of project we are engaged in when constructing a theory of function. Millikan (1989) helpfully distinguishes between descriptive definitions, theoretical definitions, and stipulative definitions. Which one are we going for? Are we engaging in something like a conceptual analysis (what she calls a "descriptive definition") of what most people have in mind when they use the term "function?" Or (as a variation on the conceptual analysis strategy) are we only trying to produce a conceptual analysis of what *modern biologists* have in mind by the term? Are we trying to produce a theoretical definition, that is, in the sense that "water = H20" constitutes a theoretical definition of water? Or are we engaged in something like pure stipulation? I will set out these three options below, and then draw the conclusion that it does not matter too much which of these options we accept, as long as we agree that our definition should be constrained by actual biological usage.

Philosophical analyses have traditionally been thought of as conceptual analyses, that is, as attempts to set out what most people have in mind, either implicitly or explicitly, when they use a term. Consider the way that philosophers have wrestled over the correct conceptual analysis of the term "knowledge." In principle, the way to discredit a conceptual analysis is pretty simple: one devises a counterexample, real or imagined, in which we would intuitively apply the term in ways forbidden by the analysis (or in which we would intuitively refrain from applying the term in ways licensed by the analysis).

If our theory of function is supposed to be a conceptual analysis in this sense, then certain popular theories would be discredited right off the bat, most obviously, the selected effects theory, which holds that the function of a trait is, roughly, the reason it evolved by natural selection. After all, there are plenty of English speakers who do not believe the theory of natural selection, or who are not familiar with the theory, but who seem to use the term "function" competently. So, the selected effects theory fails to capture what they have in mind when they use the term "function."

One could argue that the selected effects theory of function need not be a conceptual analysis of what *all* English speakers have in mind when they use the term, but only a *subset* of English speakers, namely, modern English-speaking biologists (Neander 1991). If the theory only purports to be a conceptual analysis of what modern biologists have in mind when they use the term "function," then the case for conceptual analysis is much more plausible, though some have argued that it fails even in this more limited task (see Sect. 4.3 for discussion). In the following, when I discuss the conceptual analysis strategy, I will restrict my attention to the version of this idea that holds that theories of function should be conceptual analyses of modern biological usage.

One apparent pitfall of this approach (which construes a theory of function as a conceptual analysis of modern biological usage) is that it postulates a deep conceptual divergence in ordinary usage of the term "function" that is not self-evident (e.g., Boorse 1976; Nagel 1977). For example, it implies that when an evolutionary biologist states, "the function of zebra stripes is to deter biting flies," and when a young-earth creationist states, "the function of zebra stripes is to deter biting flies," they mean something very different by the term "function." That is not self-evident. It would also imply that when the seventeenth-century English physiologist William Harvey discovered, and stated, that the function of the heart is to circulate blood, he meant something very different than what modern biologists mean when they say the same thing. That, too, is not self-evident. However, on behalf of the conceptual analyst, it should be said that concepts change over time, both in science and society, so we cannot rule out the possibility that the term "function" has undergone some change of meaning over the last several hundred years (Neander 1991). By the same token, sometimes scientists use concepts in ways that differ from how ordinary people use them.

A second possibility is that a theory of function is supposed to be something like a theoretical definition of the term "function," much like "water = H20" is a theoretical definition of the term "water." Here, if I understand it correctly, the idea is that our theory of function is supposed to identify something like a *natural kind* that underlies most correct usage of the term in question. One obvious advantage here is that, if a theory of function is supposed to be a theoretical definition of "function," it need not purport to capture what people have in mind when they use the term (scientists or otherwise), and so the sorts of objections raised against the conceptual analysis approach would simply miss their mark. Millikan (1989) is the most prominent defender of this approach, and she also accepts a form of the selected effects theory.

Note, however, that if we think that a theory of function should be a conceptual analysis of what modern biologists mean by the term "function," then the two sorts of approaches will tend to converge (Neander 1991). This is obvious from the "water = H20" example. The identity statement, "water = H20" could be seen *either* as a theoretical definition of water, *or* as a conceptual analysis of what modern chemists mean by "water." This is because scientists tend to revise their concepts over time in light of their current scientific theories.

Third, one might think that a theory of function is neither of these, but something more along the lines of a stipulative definition, that is, a recommendation for how biologists should use the term. The most important proponent of this approach to philosophical analysis generally is Carnap (1950), who referred to this as a philosophical "explication." Carnap's view is that philosophical explication always results in the *replacement* of a term or concept that is vague (which he called the "explicandum") with a new concept that is precise (the "explicatum"). It has the status of a recommendation, to be accepted or rejected on pragmatic grounds. Carnap emphasized the idea that there is often more than one acceptable and reasonable "explicatum" for one and the same "explicandum." In light of the present-day debates, Carnap probably would have accepted a form of function pluralism, which holds that there is more than one philosophical theory of function that accurately captures certain elements of biological usage and that biologists might find useful (see Sect. 5.3).

Schwartz (2004) wrote an interesting paper that applies this idea to the functions debate. He says there are some instances where there is simply no fact of the matter about whether or not a trait has a function, because biological use is too vague. As function theorists, we have to engage in a bit of stipulation. Schwartz notes, however, that we should expect substantial convergence between theoretical definitions and Carnapian-style explications. After all, "water = H20" could reasonably be understood as a Carnapian-style explication of the somewhat vague, ordinary folk notion of water.

It seems to me that the chief difference between the proponent of Carnapian-style explication and the proponent of theoretical definition is that the latter seems to require some philosophical account of what *natural kinds* are, and the proponent of Carnapian explication does not require any such account. It seems to me, moreover, that the chief difference between the proponent of Carnapian-style explication, on the one hand, and the proponent of conceptual analysis (where conceptual analysis is restricted to what modern biologists have in mind when they use a term), on the other, is that the former can give themselves more latitude for pure stipulation than the latter.

Overall, the differences between these three approaches involve fairly subtle issues about language and metaphysics. It is not entirely necessary for an aspiring function theorist to commit to one of these options in advance! What is more notable and interesting to me is the extent to which all three of these approaches share the same basic attitude about the importance of actual scientific practice in devising a theory of function. All of them agree that function concepts should reflect science well, and they should result in theories that scientists may find useful (Huneman 2013 points out that this is a good constraint on theorizing about function, even if we disagree about the precise nature of the project.). Moreover, the proponents of these various approaches have little patience with purely science-fiction counterexamples (e.g., "swampman" type cases or twin-earth examples). That is because science-fiction examples are mainly useful for evaluating theories that purport to capture ordinary ("folk") use of concepts, but none of the three meta-analytic views discussed here are trying to do that.

1.4   Overview of the Volume

The structure of this volume is very straightforward. It is a survey of the main viewpoints accompanied by critical discussion. The second chapter looks back to the state of discourse in naturalized teleology prior to the 1970s. In particular, it discusses the oft-neglected goal directedness approach to teleology that reigned in the philosophy of science from the 1940s to the 1960. It also discusses the theory of function that emerged from this tradition, namely, the goal-supporting theory of functions. The third, fourth, and fifth chapters consider, in turn, the three chief views of function that have emerged over the last 40 years. These are the selected effects theory, the fitness contribution theory, and the causal role theory. Strengths and weaknesses will be evaluated as well as more recent innovations, including the mechanistic causal role theory and the generalized selected effects theory. The sixth chapter considers three recent alternatives to these traditional theories. Most readers who have not followed the functions debate carefully from the turn of the century will likely not be familiar with them since they all emerged around that time or afterwards. These are the weak etiological theory, the systems-theoretic (or organizational) views, and the modal theory of function.

Although the volume is primarily a survey, I make no attempt to conceal my own viewpoint or to pretend to neutrality. At the most general level, my own view is that the selected effects theory captures a crucial strand of biological usage, and that all of the arguments that have been leveled against it fail. One of my goals is to explain why those arguments fail. At the same time, I think we should be pluralistic in how we understand "function." The selected effects theory captures one important strand of biological usage. It does not capture all of it. Although the volume is slanted toward the selected effects theory, I have made every effort to present alternative viewpoints carefully and fairly.

More specifically, there are three main claims I wish to advance throughout the volume, and I hope these will provoke fresh discussion and debate. I will return to these points in the closing chapter. First, I will summarize a view that I have defended in several places (Garson 2010, 2011, 2012, 2015), which I call the *generalized* selected effects theory (see Sect. 3.4). The theory states that the function of a trait in a population is that activity that led to its *differential persistence or reproduction* in that population. Unlike other forms of the selected effects theory (e.g., Neander 1983; Millikan 1984) it does not insist that the trait must have been *reproduced* in order to have a function. This theory allows, for example, unique neurological structures to have functions by virtue of how they historically contributed to their own differential persistence in that individual (that is, their persistence *over* alternative neural structures).

Second, I recommend that we accept a certain kind of pluralism about functions, where the selected effects theory and the causal role theory both capture something important about biological usage. However, I do not agree with what I take to be the dominant version of pluralism, which I call "between-discipline pluralism." This is the claim that different theories of function are appropriate to different biological sub-disciplines. For example, this type of pluralist might hold that the selected effects theory is only appropriate for (some aspects of) evolutionary biology, and that some other theory of function is most appropriate for disciplines such as physiology or molecular biology (e.g., Godfrey-Smith 1993, 200; Amundson and Lauder 1994, 446; Griffiths 2006, 3; Maclaurin and Sterelny 2008, 114; and Bouchard 2013, 86). I think this version of pluralism rests on an overly narrow conception of what the selected effects theory entails. Instead, I will advocate a view I call, "within-discipline pluralism," which emphasizes the way that different concepts of function co-exist within the same disciplines (see Sect. 5.3; also see Garson Submitted for publication b).

Third, to the extent that function statements purport to be explanatory, *in the sense of providing a causal explanation for the existence of a trait* (e.g., why zebras have stripes), I do not believe there are any contenders to the selected effects theory. That is, function statements constitute causal explanations only by virtue of citing some selected effect. This claim is contrary to claims made by proponents of the weak etiological theory (Buller 1998) and the systems-theoretic or organizational view (Schlosser 1998; McLaughlin 2001; Christensen and Bickhard 2002;

Mossio et al. 2009), who purport to capture a causal-explanatory notion of function that does not refer to selection. I will develop this point in Sects. 6.1 and 6.2. I am not claiming that explanation must always be understood in the causal sense. Rather, my point is that, to the extent that functions *are* explanatory, and to the extent that we understand explanation in the causal sense, there are no current, viable contenders to the selected effects theory.

In the following, I have chosen not to focus on artifact functions. What is it for an artifact, such as a laptop projector or a paperweight, to have a function? In order to treat this question fruitfully, we would have to delve into the way that objects are designed, manufactured, and deployed, which takes us into the philosophy of mind, the philosophy of action, and the philosophy of technology. Since I do not have the expertise to tackle those questions fruitfully, I chose to avoid them here. I have a little more to say about artifact functions in Sect. 3.2, where I discuss whether there is some intuitive notion of selection that can apply to both biological and artifact functions. I also give several references to the most recent philosophical literature for those who wish to follow it up.

One might see this as a significant lacuna in my volume. In the opinion of some, a good theory of biological functions should also help us make sense of artifact functions, too. In other words, some theorists believe that there is a single, unified notion of function that underlies talk of biological functions ("the function of eyespots on butterfly wings is to deter attack from vital organs") and talk of artifact functions ("the function of shoehorns is to help fit your heel into tight shoes"). Indeed, in the 1970s, some theorists held that an adequacy condition on any acceptable theory of biological function is that it should make sense, without further ado, of artifact functions. Treating biological functions separately from artifact functions would have struck theorists like Wright (1973) and Boorse (1976) as bizarre.

Since the early 1990s, this idea of unification has come under a lot of pressure. The reason it came under pressure is that some philosophers of biology started to think that the best account we had of biological functions—the selected effects theory—could not be applied in any straightforward way to artifacts (Neander 1991; Godfrey-Smith 1993). Instead of giving up the selected effects theory, they chose to scrap the proposed adequacy condition instead. Since my own sympathies lie with the selected effects theory, I agree with this line of thought. Some have even wondered whether the very idea of unification is a remnant of a pre-Darwinian way of thinking, in which both artifacts and biological entities derive their purposes through intelligent agency (e.g., Lewens 2004).

In the following, I endorse the idea that, as philosophers of biology, we should try to construct a theory of function that, first and foremost, makes sense of the way that biologists use the term (both their implicit and explicit commitments). This turns out to be an extremely difficult task in its own right. If that theory of function happens to make good sense of artifact functions as well, then so much the better for that theory! I will not, however, treat the idea of unification as a kind of adequacy condition on the development of the theory of function.

References

Adriens, P. R., & de Block, A. (Eds.). (2011). *Maladapting minds: Philosophy, psychiatry, and evolutionary theory*. Oxford: Oxford University Press.

American Psychiatric Association. (2013). *Diagnostic and statistical manual of mental disorders: DSM-5*. Washington, DC: American Psychiatric Association.

Amundson, R. (2000). Against normal function. *Studies in History and Philosophy of Biological and Biomedical Sciences, 31*, 33–53.

Amundson, R., & Lauder, G. V. (1994). Function without purpose: The uses of causal role function in evolutionary biology. *Biology and Philosophy, 9*, 443–469.

Bock, W. J., & von Wahlert, G. (1965). Adaptation and the form-function complex. *Evolution, 19*, 269–299.

Boorse, C. (1976). Wright on functions. *Philosophical Review, 85*, 70–86.

Boorse, C. (1977). Health as a theoretical concept. *Philosophy of Science, 44*, 542–573.

Boorse, C. (2014). A second rebuttal on health. *Journal of Medicine and Philosophy, 39*, 683–724.

Bouchard, F. (2013). How ecosystem evolution strengthens the case for function pluralism. In P. Huneman (Ed.), *Function: Selection and mechanisms* (pp. 83–95). Dordrecht: Springer.

Buller, D. J. (1998). Etiological theories of function: A geographical survey. *Biology and Philosophy, 13*, 505–527.

Burge, T. (2010). *Origins of objectivity*. Oxford: Oxford University Press.

Carnap, R. (1950). *Logical foundations of probability*. Chicago: University of Chicago Press.

Christensen, W. D., & Bickhard, M. H. (2002). The process dynamics of normative function. *The Monist, 85*, 3–28.

Cosmides, L., & Tooby, J. (1999). Toward an evolutionary taxonomy of treatable conditions. *Journal of Abnormal Psychology, 108*, 453–464.

Crane, T. (2003). *The mechanical mind* (2nd ed.). London: Routledge.

Craver, C. (2001). Role functions, mechanisms, and hierarchy. *Philosophy of Science, 68*, 53–74.

Craver, C. (2013). Functions and mechanisms: A perspectivalist view. In P. Huneman (Ed.), *Function: Selection and mechanisms* (pp. 133–158). Dordrecht: Springer.

Cummins, R. (1975). Functional analysis. *Journal of Philosophy, 72*, 741–765.

Daniels, N. (1985). *Just health care*. New York: Cambridge University Press.

Davies, P. S. (2000). Malfunctions. *Biology and Philosophy, 15*, 19–38.

Davies, P. S. (2001). *Norms of nature: Naturalism and the nature of functions*. Cambridge, MA: MIT Press.

Davies, P. S. (2009). Conceptual conservatism: The case of normative functions. In U. Krohs & P. Kroes (Eds.), *Functions in biological and artificial worlds* (pp. 127–146). Cambridge, MA: MIT Press.

Doolittle, W. F. (2013). Is junk DNA bunk? A critique of ENCODE. *Proceedings of the National Academy of Sciences*. doi:10.1073/pnas.1221376110

Elliot, R. (1982). Faking nature. *Inquiry, 25*, 81–93.

ENCODE Project Consortium. (2012). An integrated encyclopedia of DNA elements in the human genome. *Nature, 489*, 57–74.

Faucher, L., & Forest, D. (Eds.). (Submitted for publication). *Defining mental disorder: Jerome Wakefield and his critics*. Cambridge, MA: MIT Press.

First, M. B. (2007). Potential implications of the harmful dysfunction analysis for the development of DSM-V and ICD 11. *World Psychiatry, 6*, 158–159.

Garson, J. (2010). Schizophrenia and the dysfunctional brain. *Journal of Cognitive Science, 11*, 215–246.

Garson, J. (2011). Selected effects functions and causal role functions in the brain: The case for an etiological approach to neuroscience. *Biology and Philosophy, 26*, 547–565.

Garson, J. (2012). Function, selection, and construction in the brain. *Synthese, 189*, 451–481.

Garson, J. (2013). The functional sense of mechanism. *Philosophy of Science, 80*, 317–333.

Garson, J. (2015). *The biological mind: A philosophical introduction*. London: Routledge.

Garson, J. (Submitted for publication a). Ecological restoration and biodiversity conservation. In J. Garson, A. Plutynski., & S. Sarkar (Eds.), *Routledge handbook for philosophy of biodiversity*. London: Routledge.

Garson, J. (Submitted for publication b). How to be a function pluralist. *British Journal for the Philosophy of Science*.

Garson, J., & Piccinini, G. (2014). Functions must be performed at appropriate rates in appropriate situations. *British Journal for the Philosophy of Science, 65*, 1–20.

Godfrey-Smith, P. (1993). Functions: Consensus without unity. *Pacific Philosophical Quarterly, 74*, 196–208.

Godfrey-Smith, P. (2007). Innateness and genetic information. In P. Carruthers, S. Lawrence, & S. Stich (Eds.), *The innate mind, vol. 3: Foundations and the future* (pp. 55–68). Oxford: Oxford University Press.

Graur, D., et al. (2013). On the immortality of television sets: "Function" in the human genome according to the evolution-free gospel of ENCODE. *Genome Biology and Evolution, 5*, 504–513.

Griffiths, P. E. (2006). Function, homology, and character individuation. *Philosophy of Science, 73*, 1–25.

Hausman, D. (2011). Is an overdose of paracetamol bad for one's health? *British Journal for the Philosophy of Science, 62*, 657–668.

Huneman, P. (2013). Introduction. In P. Huneman (Ed.), *Function: Selection and mechanisms* (pp. 1–17). Dordrecht: Springer.

Katz, E. (1992). The big lie: Human restoration of nature. *Research in Philosophy and Technology, 12*, 231–241.

Kellis, M., et al. (2014). Reply to Brunet and Doolittle: Both selected effect and causal role elements can influence human biology and disease. In *Proceedings of the National Academy of Sciences of the United States of America* (Vol. 111, p. E3366).

Kingma, E. (2010). Paracetamol, poison, and polio: Why Boorse's account of function fails to distinguish health and disease. *British Journal for the Philosophy of Science, 61*, 241–264.

Kitcher, P. (1993). Function and design. *Midwest Studies in Philosophy, 18*, 379–397.

Lewens, T. (2004). *Organisms and artifacts: Design in nature and elsewhere*. Cambridge, MA: MIT Press.

Lo, Y. S. (1999). Natural and artifactual: Restored nature as subject. *Environmental Ethics, 21*, 247–266.

Macdonald, G., & Papineau, D. (Eds.). (2006). *Teleosemantics*. Oxford: Clarendon Press.

Maclaurin, J., & Sterelny, K. (2008). *What is biodiversity?*. Chicago: University of Chicago Press.

Maynard Smith, J. (2000). The concept of information in biology. *Philosophy of Science, 67*, 177–194.

McLaughlin, P. (2001). *What functions explain: Functional explanation and self-reproducing systems*. Cambridge: Cambridge University Press.

Mebius, A. (2014). A weakened mechanism is still a mechanism: On the causal role of absences in mechanistic explanation. *Studies in History and Philosophy of Biological and Biomedical Sciences, 45*, 43–48.

Millikan, R. G. (1984). *Language, thought, and other biological categories*. Cambridge, MA: MIT Press.

Millikan, R. G. (1989). In defense of proper functions. *Philosophy of Science, 56*, 288–302.

Millikan, R. (2004). *Varieties of meaning*. Cambridge, MA: MIT Press.

Moghaddam-Taaheri, S. (2011). Understanding pathology in the context of physiological mechanisms: The practicality of a broken-normal view. *Biology and Philosophy, 26*, 603–611.

Moss, L. (2012). Is the philosophy of mechanism philosophy enough? *Studies in the History and Philosophy of Biological and Biomedical Sciences, 43*, 164–172.

Mossio, M., Saborido, C., & Moreno, A. (2009). An organizational account for biological functions. *British Journal for the Philosophy of Science, 60*, 813–841.

Nagel, E. (1961). *The structure of science*. New York: Harcourt, Brace and World.

Nagel, E. (1977). Teleology revisited: Goal directed processes in biology and functional explanation in biology. *Journal of Philosophy, 74*, 261–301.

Nanay, B. (2010). A modal theory of function. *Journal of Philosophy, 107*, 412–431.

Neander, K. (1983). *Abnormal Psychobiology*. Dissertation, La Trobe.

Neander, K. (1991). Functions as selected effects: The conceptual analyst's defense. *Philosophy of Science, 58*, 168–184.

Neander, K. (2002). Types of traits: The importance of functional homologues. In A. Ariew, R. Cummins, & M. Perlman (Eds.), *Functions: New essays in the philosophy of psychology and biology* (pp. 390–415). Oxford: Oxford University Press.

Neander, K. (2012). Teleosemantic theories of mental content. Stanford Encyclopedia of philosophy. http://plato.stanford.edu/entries/content-teleological/

Neander, K., & Rosenberg, A. (2012). Solving the circularity problem for functions. *Journal of Philosophy, 109*, 613–622.

Nervi, M. (2010). Mechanisms, malfunctions, and explanations in medicine. *Biology and Philosophy, 25*, 215–228.

Nesse, R. M. (2007). Evolution is the scientific foundation for diagnosis: Psychiatry should use it. *World Psychiatry, 6*, 160–161.

Rosenberg, A., & Neander, K. (2009). Are homologies (selected effect or causal role) function free? *Philosophy of Science, 76*, 307–334.

Ryder, D. (Submitted for publication). *Models of the brain: Naturalizing human intentionality*. Oxford: Oxford University Press.

Sarkar, S. (2000). Information in genetics and developmental biology: Comments on Maynard Smith. *Philosophy of Science, 67*, 208–213.

Sarkar, S. (2013). Information in animal communication: When and why does it matter? In U. Stegmann (Ed.), *Animal communication theory: Information and influence* (pp. 189–205). Cambridge: Cambridge University Press.

Schlosser, G. (1998). Self-re-production and functionality: A systems-theoretical approach to teleological explanation. *Synthese, 116*, 303–354.

Schwartz, P. H. (2004). An alternative to conceptual analysis in the function debate. *The Monist, 87*, 136–153.

Schwartz, P. H. (2007). Defining dysfunction: Natural selection, design, and drawing a line. *Philosophy of Science, 74*, 364–385.

Shea, N. (2013). Inherited representations are read in development. *British Journal for the Philosophy of Science, 64*, 1–31.

Siipi, H. (2003). Artefacts and living artefacts. *Environmental Values, 12*, 413–430.

Stegmann, U. (2009). A consumer-based teleosemantics for animal signals. *Philosophy of Science, 76*, 864–875.

Sterelny, K. (2000). The 'genetic program' program: A commentary on Maynard Smith on information in biology. *Philosophy of Science, 67*, 195–201.

Wakefield, J. C. (1992). The concept of mental disorder: On the boundary between biological facts and social values. *American Psychologist, 47*, 373–388.

Walsh, D. M. (1996). Fitness and function. *British Journal for the Philosophy of Science, 47*, 553–574.

Walsh, D. M., & Ariew, A. (1996). A taxonomy of functions. *Canadian Journal of Philosophy, 26*, 493–514.

Wright, L. (1973). Functions. *Philosophical Review, 82*, 139–168.

Chapter 2
Goals and Functions

Abstract Contemporary philosophical debates about biological function started in the early 1970s, and they originated from earlier, related, debates about the nature of goal directed systems. These discussions were rooted in scientific advances in the 1920s and 1930s pertaining to cybernetic machines and homeostatic systems, which appear to be purposeful or goal-directed despite not having any conscious intentions. By the 1950s, there were two major philosophical traditions for analyzing goal directedness, the behavioristic and the mechanistic. According to the behavioristic approach, favored by theorists like Gerd Sommerhoff and Richard Braithwaite, a goal directed system is one that exhibits plasticity and persistence in its outward behavior. According to the mechanistic tradition, favored by Ernest Nagel and Norbert Wiener, a goal directed system must be governed by the right sort of mechanism (such as negative feedback). Both of those traditions faced severe philosophical criticism in the 1960s and 1970s. I begin this chapter by sketching the historical background of the earlier debates about goal directedness. I then present the behavioristic analyses of Sommerhoff and Braithwaite, and enumerate several serious criticisms. I discuss mechanistic approaches, namely those of Nagel and the cyberneticists, and their critics.

Keywords Goal directedness · Naturalized teleology · Cybernetics · Homeostatic systems · Negative feedback

2.1 The Theory of Goal Directed Systems

A beaver builds a dam in order to surround itself with water. A frog orients its head to catch a buzzing fly. An anaerobic bacterium swims toward geomagnetic north to move away from oxygen-rich water. Some plants track the movement of the sun across the sky (heliotropic behavior). One of the most distinctive characteristics of living things, and perhaps even life's defining characteristic, is its apparent goal directedness. We explain why living things act the way they do in terms of certain states that those behaviors tend to bring about.

© The Author(s) 2016
J. Garson, *A Critical Overview of Biological Functions*,
Philosophy of Science, DOI 10.1007/978-3-319-32020-5_2

Traditionally, this sort of explanation has been called a "teleological" explanation. A teleological explanation is one that purports to explain the existence of an entity (such as an organism, trait, or behavior), in terms of some effect the entity tends to bring about. Why do we have teeth? Because teeth help us chew food. Why is the cat crouching by the mouse hole? Because crouching by a mouse hole is a good way to catch mice. Terms like "purpose," "end," and "goal," are signals for teleological explanations. Organisms, their parts, and their behaviors, seem particularly apt for such teleological explanations. At the beginning of Sect. 3.1, I explain more carefully what teleological explanations are.

What do functions have to do with goals? Goals are different from functions. Something can have a goal without having a function. People and other organisms have goals, but they do not have functions. When we consider something as a self-contained entity—an uncontained container—we typically do not attribute a function to it, but we do attribute goals to it. By the same token, something can have a function without having any goals. An artifact such as a paperweight has a function, which is derived from the intention of the person who made it, or perhaps by the way it is currently used. But it does not have a goal.

If goals are different from functions, why talk about goals here? This is a volume on functions, not goals. There are three good reasons to talk about goals. First, although they are very different, there is an intimate relationship between them. Goals and functions both have a role in teleological explanations. To explain an organism's behavior in terms of a goal is to explain it in terms of some state it tends to promote. Similarly, according to one long tradition of thinking about functions (see Sect. 3.1), to attribute a function to an item is to explain that item in terms of some effect it creates. Goal-explanations and function-explanations are two species of the same genus, like apples and oranges are two kinds of fruit.

But the relationship is potentially tighter than that. This brings us to a second reason for talking about goals here. Some theorists have tried to *define* functions in terms of goals. On the surface, this is a tempting idea. When we say that the heart has the function of circulating blood, what more are we saying than that blood circulation promotes one of the organism's goals, namely to survive? And when we say that a paperweight has the function of holding down papers, what more are we saying than that holding down papers promotes the goals of the people who make and use it? Philosophers who developed thoughts along these lines include Nagel 1953, 1961, 1977; Wimsatt 1972; Boorse 1976; Adams 1979; Schaffner 1993. Recent theorists who have explored the notion of goal-directedness and its importance for functions include McShea (2012, 2013); Trestman (2012); Piccinini (2015, Chap. 6); Maley and Piccinini (Submitted for publication). Although these authors disagree about what exactly goals are, they all agree with the idea that functions are contributions to goals. (Interestingly, and as I will note in some detail in Sect. 4.1, Boorse's view is often described as a fitness-contribution view, but this is a mistake. Boorse accepted the basic account of goal-directedness provided by Nagel and Schaffner, where goals are defined in cybernetic terms. He just thought that, *in the context of physiology and biomedicine*, goals are simply tantamount to fitness-contributions.)

A third reason for talking about goals here is that the theory of goal directed systems provided an important historical backdrop for modern discussions of biological function. The notion of goal directedness represented the first sustained attempt by philosophers and scientists to make sense of teleological explanations in natural terms. That is, it represented an attempt to understand how scientists could legitimately use and apply terms like "goal," "purpose," and "end" to biological systems (and even machines) without assuming that those systems were created by God for special purposes, or that those animals, plants, and machines, had thoughts and intentions of their very own.

I will begin with the historical context, and then dive into the conceptual part. This is not meant to be a detailed historical exposition. Instead, it is an attempt to illuminate in very broad strokes the social and scientific context that motivated this thinking. Historically, philosophical discussion of goal directedness originated from two different traditions in early twentieth century science (though I suppose that historians can trace it back much further). The first tradition was organismic biology (see, e.g., Gilbert and Sarkar 2000 for discussion of this tradition). There were several biologists in the first half of the century that rejected the vitalism of people like Hans Driesch, but still insisted that there was some fundamental difference between living systems and purely physical systems. What is distinctive about living systems, in their view, is that they are goal directed and purposeful (Rignano 1931; Russell 1945; Bertalanffy 1950). So, they asked, what makes those systems purposeful?

The second tradition was cybernetics, which flourished during World War II and was devoted to the study and manufacture of "servomechanisms," or life-like machines (Wiener 1948). A heat-seeking missile is an example. It pursues its object with a deadly accuracy that we are tempted to describe as "purposeful." There was an important paper published in *Philosophy of Science* in 1943 that attempted to justify the idea that such machines are inherently purposeful. This was written by the American mathematician Norbert Wiener, the Mexican physiologist Arturo Rosenblueth (who had earlier worked with Walter Cannon), and the American engineer Julian Bigelow (Rosenblueth et al. 1943). They argued that the sciences required a concept of purposiveness that could apply to machines as well as organisms, and that did not appeal to conscious intentions or design. A heat-seeking missile, they argued, has the goal or purpose of striking its target. Moreover, this goal is intrinsic to it, in the sense that having that goal does not logically depend on the existence of human intentions. Philosophers sympathetic to this tradition quickly extended this notion of intrinsic purposiveness to describe certain biological systems as well, such as homeostatic mechanisms, e.g., the system in mammals that regulates the water content of the blood (e.g., Nagel 1953).

So, what do we mean when we say that a system is goal directed? Let's approach this, first, from the standpoint of conceptual analysis of ordinary language. (Perhaps we will choose to depart from the confines of conceptual analysis later.) One natural idea is that *purposeful human behavior* is the best starting point for thinking about goals (Taylor 1950; Woodfield 1976). To say that a person has a goal is usually just to say that the person has a desire or an intention. To say I have the goal of visiting

France is just to say that is what I intend to do. When we say that animals have goals we are also attributing intentions to them. Perhaps their intentions are not as sophisticated as ours, and perhaps they are not expressed in the same sort of linguistic or syntactical structures as ours, but they are intentions nonetheless.

This sort of analysis, which takes as its starting point conscious human intentions, has an obvious limitation. It would not apply to machines such as homing torpedoes. Nor would it apply to the behavior of creatures that do not have minds, such as plants or insects. But we often want to say that their behavior is goal directed, too. The proponent of the idea that goals are conscious intentions will insist that, when we say that plants are goal directed, or that machines are goal directed, we are speaking metaphorically. We are engaged in a bit of anthropomorphism. We observe that their behavior is *in some respects* like conscious intentional behavior, and so we metaphorically extend the notion of goals to describe them (Woodfield 1976, 194; Nissen 1993, 48).

I do not see any easy way to refute a philosopher who insists that all talk of goal directedness presupposes conscious intentions. But does that mean that there is nothing more to be said about the notion of goal directedness? Not at all. For notice that the proponent of this analysis (goals as conscious intentions) is willing to concede that the behavior of some organisms and machines is similar *in some respects* to human purposeful behavior, and that it is *by virtue of* these similarities that we are inclined to call those machines or organisms "purposeful." So, *in what respect*, precisely, is the behavior of machines and insects similar to conscious human behavior? Even if we believe that goals are *conceptually* equivalent to intentions, we can still discover some interesting things about the behaviors that are distinctive of goal directed systems.

Many philosophers have pursued this way of thinking. There are two lines of thought here. The first line of thought is that goal directed systems can be identified by a distinctive pattern of behavior. In other words, the idea is that, just by looking at a system from the outside, in complete ignorance of its inner mechanisms, we can judge that the system is goal directed. The core idea here is that a goal directed system is one that has the right kind of behavioral flexibility (often dubbed "persistence and plasticity") in attaining or maintaining the presumed goals. I will call this the "behavioristic" approach to goal directedness.

The second line of thought is that, when we say that a system is goal directed, we are not *merely* saying something about its distinctive patterns of behavior. Rather (or in addition to that) we are saying something about the mechanism that gives rise to that behavior. I will refer to these as "mechanistic" analyses of goal directedness. One version of this idea, which goes back to the cyberneticists, is that goal directed behavior is behavior that is governed by a special sort of mechanism called a "negative feedback system." Things get somewhat confusing here because, although cyberneticists like Wiener *claimed* to offer a purely behaviorist analysis of goal directedness, they clearly were not. They were tacitly, if not explicitly, committed to the claim that what makes a system goal directed is that it has the right sort of inner mechanism (see Wimsatt 1971 for discussion). Moreover, even theorists such as Sommerhoff, who were trying to articulate a purely behavioristic theory of goal directedness, were forced to make certain assumptions about underlying mechanisms. In Sect. 2.3, I will describe this mechanistic approach and identify its shortcomings.

2.2   From Intentions to Behavior

For the time being, let's set aside the idea that having a goal is simply tantamount to having conscious intentions. Let's suppose there is a literal sense in which animals, plants, and machines, can exhibit goal directedness even if they do not all have intentions. One natural idea here is that goal directedness has something to do with their characteristic forms of behavior. A popular slogan in the literature is that goal directed systems exhibit persistence and plasticity (e.g., Nagel 1977, 272).

Consider a heat-seeking missile closing in on a moving target. In what respect is it life-like? First, in the ideal case, if the target moves, the missile moves, too. The missile adjusts its trajectory to match the behavior of the target. If something interferes, such as a strong gust of wind, then (again, in the ideal case) the missile gets back on track. It exhibits *persistence*: it tends toward the goal even in the face of obstacles. This is what makes a heat-seeking missile different from, say, a projectile launched from a catapult. If one launches a projectile from a catapult and the target moves, the projectile does not change its course. Persistence also characterizes much of human purposeful behavior.

The second, plasticity, is the idea that one and the same end can be achieved through a variety of different starting points. The missile clearly exhibits plasticity in this sense. There is wide range of points from which the missile can be launched, and it will still result in the same outcome. In short, plasticity has more to do with the starting point of the goal directed behavior. Persistence has more to do with the behavior as it is being carried out.

When we attribute plasticity and persistence to an item, it seems that we are merely describing its outward behavior, independent of the mechanism by which it achieves this remarkable effect. It may be the case, as a matter of contingent empirical fact, that all systems that exhibit plasticity and persistence have certain inner mechanisms in common, such as negative feedback mechanisms. But arguably, when we judge that a system exhibits plasticity and persistence, we can remain neutral about the sort of inner mechanisms that cause the behavior. (Trestman 2012, 208 emphasizes this point in developing an approach to goal directedness.)

The first important attempt to give a purely behavioral analysis of goal directedness is due to the biologist Sommerhoff (1950; also see his 1969 and 1974). (One might object that Rosenblueth et al. 1943 have precedence here, but for reasons that I will shortly describe, their view is a version of the mechanistic approach.) Sommerhoff was very liberal in his use of neologisms, and he used fairly dense mathematical formalisms, so his work is not easy to read, but the basic ideas are fairly simple. Any goal directed system involves three features: the system itself (S), the target of the system, which is some object in its environment (E), and a goal (G). Consider a shooter who is aiming a rifle at a target. As the target moves around, the shooter repositions the rifle to match the target. Here, the shooter is the goal directed system (S). The target (E) is the thing that the shooter is aiming at. It is a physical object that moves somewhat independently of S. Finally, S has the goal G of hitting E. G is the outcome that explains S's behavior. The same three variables can describe a chick pecking at some grain. The chick is the goal directed system, S. The grain is the relevant object in the environment, E. The goal, G, is that the chick eats the grain.

How would we model such a system? The most obvious way is by the use of two variables, each of which can take on different numeric values. We will call them V_S and V_E. V_S is a variable that describes some property of S. In the shooter example, we will use V_S to designate the angle of the rifle. Second, there is a variable V_E that represents some property of E. In the shooter example, we can let V_E represent the angle of the target. Finally, let's assume that the shooter's goal is simply to keep her sights on the target, rather than to destroy it. So we can describe the goal of the system, G, in terms of a certain relationship between V_S and V_E, namely one of equivalence. In other words, the shooter's goal is *that* $V_S = V_E$.

I still have not stated what it means to say that a system is goal directed, that is, what it means to say, "S has G with respect to E." Sommerhoff's main insight here is that, when we say that a system is goal directed, we are not merely describing the actual behavior of the system, here and now. We are making a counterfactual claim. We are describing how the system variable *would* change if the environmental variable were to change in various ways. Specifically, we are saying that, if V_E changed in various ways (and within a given range), V_S would change as well, *in such a way that G is satisfied* (Sommerhoff 1950, 54). We can summarize his rather abstract discussion as follows:

> Where S is a system, E is an object in S's environment, Vs is a variable that represents S, V_E is a variable that represents E, and G is some relationship between Vs and V_E, "S is goal directed with respect to E and G" means:

(i) There exists a range of different values of V_S and V_E such that, if V_E were to change, V_S would also change in such a way that G is satisfied.

(I will add a second condition momentarily.) Much more could be added in order to extend Sommerhoff's analysis to more complex cases. Sommerhoff's examples typically embody what Canfield (1966, 5) calls the "target" schema rather than the "furnace" schema. In the target schema (such as a heat-seeking missile) a system is trying to attain some objective, rather than to maintain some inner state (such as temperature homeostasis). It is impossible to do justice to the full complexity of Sommerhoff's analysis, but I think (i) represents his core insight.

Most of the examples that Sommerhoff uses (the rifle example, the pecking example) involve a system adapting its behavior to a target that behaves in a way that is more or less independent of S. In other words, the relationship between S and E is asymmetrical: if S is goal directed with respect to E, E is typically not goal directed with respect to S (at least not with respect to the same goal). V_S depends on V_E in an asymmetrical matter. I will include this condition explicitly into the definition (though Sommerhoff (1950, 61) also mentions it):

(ii) V_S is asymmetrically dependent on V_E.

Here is a problem that arises for Sommerhoff's analysis, and it is one that affects it in its very core. I call it the problem of overbreadth. A purely behavioral criterion of goal directedness cannot distinguish goal directed systems from those that intuitively lack it. Consider a marble rolling to the bottom of a glass bowl, or a stretched rubber band snapping back to its original configuration. From a purely behavioral standpoint, either system exhibits plasticity (it can reach the same end point from a variety of starting points) and persistence (it can adjust its trajectory in the face of obstacles). But they are not goal directed. To call such systems "goal directed" would seem to trivialize the very notion.

I think the only way to avoid this problem is to give up a purely behavioral analysis. In other words, when we say that a system is goal directed, we are saying something about the mechanism that gives rise to the behavior, rather than (or in addition to) the behavior itself. A marble rolling down the side of a bowl is not goal directed because the mechanism that governs the marble's behavior does not have the right sort of complexity. The marble is too internally homogenous. Rather, when we say of, for example, a missile, that it is goal directed, we are implying that there are a number of different internal components (inside of the missile) that function somewhat independently of one another, and that they have to work together in just the right way in order to achieve the outcome (though all of this needs to be spelled out). Sommerhoff discusses this idea in some detail. He says that in order for a system to be goal directed, its governing components have to be "epistemically independent" of one another. I will return to this condition in the next section because it represents a kind of mechanistic orientation.

A somewhat different analysis of goal directedness comes from Braithwaite (1953). As I read him, Braithwaite emphasizes the idea that one and the same goal can be achieved in many different environments. That is, when we say that a system is goal directed, what we are saying is that the system tends to achieve the goal in large number of different circumstances. A post office in New York City bears the following inscription: "Neither snow nor rain nor heat nor gloom of night stays these couriers from the swift completion of their appointed rounds." For Braithwaite, this is the essence of goal directedness: "...the essential feature, as I see it, about plasticity of behavior, is that the goal can be attained under a variety of circumstances, not that it can be attained by a variety of means" (331–332).

Suppose there is a system S in some internal state i. Let S be a rat and i be the state of hunger. Suppose we put a bit of food in the room with it. Then S has the goal G, namely, the goal of eating the food. What do we mean when we say that S in i has goal G (the hungry rat has the goal of eating food)? We mean that there are a large number of different environmental conditions under which the rat will obtain its goal. Braithwaite's formulation, like Sommerhoff's, is fairly technical. To be precise, Braithwaite says that a system is goal directed with respect to G when the cardinality ("variancy") of the set of field conditions that, together with i, will result in a G-achieving causal chain, is greater than one (330–331). One nice feature of Brathwaite's analysis is that goal directedness is a property that comes in degrees. One might think that the larger the number of field conditions under which the system can achieve G, the greater the degree of goal directedness it possesses.

Braithwaite's analysis faces a number of problems. First, there is a problem of grain (Woodfield 1976, 34–35 makes this point). The number of field conditions under which a system can achieve its goal partly depends on how finely we discriminate between those conditions. Suppose that I am eating a sandwich in my office. Suppose that the precise configuration of air molecules in my office at this very moment represents one field condition, and its precise configuration two seconds from now represents another. Then my eating the sandwich is goal directed *merely* because I manage to eat my sandwich under a vast number of different field conditions.

I do not think this is an insoluble problem but this is the sort of thing that Braithwaite's analysis will have to register. Presumably we would want to say that two field conditions are different only insofar as they have different effects on the organism, such that the organism must adjust its behavior somewhat to achieve the same end. But then the analysis seems to have much more to do with the inherent adaptability of the system, rather than the number of environmental conditions. In other words, it seems to me that Braithwaite's analysis has the wrong focus, because it defines goal directedness by looking outward to the number of environmental conditions, rather than inward to the internal configuration of the system.

Second, Braithwaite's analysis is susceptible to what Scheffler (1959) called the "problem of multiple goals." Here is one version of the problem. Suppose there are many field conditions that, together with the internal state of the system, result in a certain event G (the presumptive goal). Suppose that every causal chain that results in G has a further consequence, namely H (for example, every time the rat eats a piece of cheese, it defecates). Then Braithwaite's analysis would force us to say that H is also a goal. That seems counterintuitive.

Third, Braithwaite's analysis shares, with Sommerhoff's, the problem of overbreadth, though unlike Sommerhoff, I am not aware that Braithwaite made any attempt to resolve it. The problem is that his simple schema can apply to ersatz goal directedness, such as rolling marbles and swinging pendulums, as well as genuine cases. Even if these behavioral or environmental analyses represent necessary conditions on goal directedness, they are not sufficient. That suggests that we have to supplement our behavioral analysis with some sort of mechanistic analysis. We must turn our focus inward, to the nature of the system that generates the behavior, rather than the behavior itself or its environment.

2.3   From Behavior to Mechanisms

We have seen in the last section how purely behavioral analyses of goal direct-edness stumble, and they stumble because of the problem of overbreadth. Theorists like Andrew Woodfield (1976) use the failure of behavioral analyses to support their own mentalistic theories of goal directedness. As I noted in Sect. 2.1, Woodfield's view is that, as a piece of conceptual analysis, goals are simply con-scious intentions (also see Nissen 1993). To attribute a goal to a system is to attribute an intention to it. In his view, plasticity and persistence are, at best, *reliable indicators* of the presence of conscious intentions, but they are fallible indicators, in the way that crying is a reliable, though fallible, indicator of sadness. In Woodfield's view, purely naturalistic approaches to goal directedness inevitably go wrong because they mistake a reliable indicator of goal directedness (plasticity and persistence) for goal directedness itself. As I noted above, I do not think there is any easy way to refute this point. The best way to respond to Woodfield's point is to come up with a good naturalistic analysis of goal directedness that avoids the problem of overbreadth.

The most obvious idea for avoiding the problem of overbreadth is to take a more mechanistic approach. In other words, in order for a system to be goal directed, it is not enough that it behaves in a certain way. Rather, the behavior has to be caused in the right way. The behavior has to be caused by the right sort of inner mechanism. The reason we do not attribute goal directedness to a rolling marble is that its inner constitution is too simple. It does not have enough moving parts, or the right sort of organization between those moving parts. It is just too internally homogenous. On the other hand, a moth that instinctively heads for the light has the right sort of inner constitution to qualify as goal directed. It has a number of moving parts that have to work together in just the right way to make its phototropic behavior possible.

So, what exactly must this inner constitution be like in order for the system to qualify as goal directed? There are two ideas here. The first is that the system has to be governed by negative feedback. The second is that the components of the system must have the right sort of independence. Both approaches are problematic. I will begin with negative feedback.

In their important paper, Rosenbleuth et al. (1943) attempt to explicate a natu-ralistic notion of purposefulness that applies to machines as well as animals. Unfortunately, they are not entirely consistent in their terminology. Sometimes they use "purpose" very generally: "the term purposeful is meant to denote that the act or behavior may be interpreted as directed to the attainment of a goal" (18). Sometimes they say that purposeful behavior is just behavior controlled by negative feedback: "all purposeful behavior may be considered to require negative feed-back" (19). I think the most consistent way to interpret their view is that goal directed systems are, necessarily, governed by negative feedback mechanisms. So what is negative feedback? As it turns out, defining the notion of negative feedback raises a host of fresh problems.

A feedback system is a special kind of input-output system. It takes inputs from its environment and produces certain outputs as a result. (There has to be some objective way of demarcating what constitutes its input and what constitutes its output. If I kick a pile of rocks and cause them to scatter, we would not say that the force of my kick is the input and the scattered rocks are output. I will not attempt to define these notions.) More specifically, a feedback system takes some of the energy from its output and uses it as input, making a loop. A thermostat is a kind of feedback mechanism. It detects the temperature of the room (its input) and then uses that information to produce a certain output (e.g., switching on a furnace). When the temperature of the room increases (as a result of its behavior) it does something else (switches it off). So, at the very minimum, a feedback system has to be equipped with a sensor (that monitors some feature of the environment) and an effector (that produces a behavior). Specifically, the effector has to be able to produce a behavior that modifies the environment in a way that the sensor can detect. That is what creates the feedback loop.

There are two sorts of feedback loops, positive and negative. In a positive feedback loop, some of the energy of the output feeds back into the system and produces more of the same sort of behavior. It has an amplifying effect. In a negative feedback loop, some of the energy of the output feeds back into the system and produces the opposite sort of behavior. It has a dampening effect. Negative feedback devices are common in the natural world because they give rise to self-regulating behavior. For example, negative feedback is used to regulate the level of synaptic dopamine produced by mesolimbic dopamine neurons (see Kandel et al. 2013). When those neurons are activated, they release some dopamine into the synapse. The axon terminal of the neuron has a number of autoreceptors that monitor the amount of dopamine in the synapse. When the level gets too high (that is, when a large number of autoreceptors are bound by dopamine), the autoreceptor sends a signal to stop releasing dopamine. When the level gets too low, it sends a signal to release more. The dopamine neuron constantly monitors its own behavior in order to modify that behavior, and it uses negative feedback to do so.

Negative feedback systems can give rise to plasticity and persistence. Suppose we say that the goal of the dopamine neuron is to maintain a constant level of synaptic dopamine. The neuron exhibits persistence in maintaining this goal. Various perturbations cause the level to increase or decrease, and the neuron modifies its own behavior in such a way as to maintain a steady level. It also exhibits plasticity, as it can achieve this effect across a wide range of starting points. Heat-seeking missiles do the same sort of thing. The missile senses the direction of its target. It then uses this information to modify its own trajectory. This is achieved through an error-minimization computation. It calculates the difference (error) between its own trajectory and the trajectory of the target. It then adjusts its direction in such a way as to minimize that difference.

The reason that the rolling marble is not a goal directed system is that it does not have the right inner constitution. It does not have anything that would objectively correspond to a distinctive sensor and effector. Nor is there anything in its inner constitution that would instantiate an error-minimization computation. The marble is completely passive in relation to its environment. Its behavior is most effectively explained in terms of the external forces (gravity and friction) that control it.

There are several major problems with this feedback approach. I will discuss three of them. The first has to do with the definition of "feedback." Wimsatt (1971) argued that the term is poorly defined. For example, Rosenblueth et al. (1943, 19) define a feedback system simply as one in which "some of the output energy of the apparatus or machine is returned as input." But this general definition encounters trivial counterexamples. Wimsatt (1971, 251) claims that any "closed loop of material transport" constitutes a feedback system, for example, two hoses that cycle water around and around.

Of course, nothing prevents us from developing a much richer definition of feedback (as in Adams 1979). One could say that a negative feedback system must have some mechanism by which it receives signals emanating from the presumptive target, and it must use this information to modify its trajectory. Presumably, this will involve an error-minimization computation (that is, it must have an internal representation of the target, and of its current behavior, and it must be able to compute the difference between them and use this information to adjust its behavior). But the problem is that we are invoking several terms that stand in as much need of clarification as the term "goal directedness" itself, such as "information," "computation," and "representation." Short of an analysis of these crucial notions, it is hard to tell whether this constitutes an advance.

Second, even if we could avoid the definitional problems of negative feedback, it is not clear that having a negative feedback device (in some suitably rich sense) is either necessary or sufficient for goal directedness. Woodfield (1976) summarizes the problems well. He points out that many electronic devices, such as televisions and radios, utilize various electronic feedback mechanisms, but they are not thereby goal directed (189). So, being controlled, in part, by negative feedback is not sufficient for goal directedness. Nor is it necessary. Consider the frog that snaps at a passing fly. Once the tongue-snap is triggered, it is no longer controlled by negative feedback. If the fly moves, the frog cannot modify the trajectory of its tongue in the course of a single snap. But the tongue-snap itself, that is, that specific piece of behavior, seems goal directed (191). See Faber (1986, 80), however, who criticizes Woodfield's choice of examples. Faber points out that some of the feedback systems in electronic equipment, such as feedback oscillators, are not negative feedback devices, and that, even though a single tongue-snap is not governed by negative feedback, it is part of a system of behavior that is, so it still fits the general analysis.

Third, and finally, even if we have a satisfying and non-vacuous notion of negative feedback, we encounter the problem of the missing goal object (Scheffler 1959). The basic idea behind the negative feedback approach is that in order for a system to be goal directed it must utilize signals from the target to modify its behavior. Yet presumably, a system can be goal directed even if the target does not exist. People can have goals regarding imaginary objects, such as the fountain of youth. A cat can crouch by a hole with the goal of catching a mouse, even if there are no mice around. The cat is not utilizing signals from the target to guide its behavior. (Ehring 1984, 501 elaborates a related problem for Nagel's account.)

Let's set aside the negative feedback approach. A second approach emphasizes the idea that, in a goal directed system, the controlling variables are all independent of, or orthogonal to, one another, in a special sort of way (Sommerhoff 1950; Nagel 1953, 1961, 1977). Sommerhoff called this the "epistemic independence" condition. Ernest Nagel also developed this idea. I will first explain Nagel's own theory of goal directedness very briefly, and then I will return to this notion of independence.

Nagel proposed something like a mechanistic account of goal directedness. As I noted above, for Sommerhoff, the paradigmatic cases of goal directedness involve a system attempting to attain some end, such as a shooter aiming for a target or a bird pecking at grain. In contrast, Nagel adopted what Canfield (1966, 5) describes as a "furnace" schema. Nagel was mainly interested in homeostatic behavior, such as the system that maintains a relatively stable body temperature in the face of external changes, or the system that ensures that the water content of the blood stays around 90 % regardless of fluctuations in intake. Here, goal directedness is not so much about achieving a certain end as it is about maintaining a steady internal state. The environment is, if anything, a source of perturbation. I think of it as a mechanistic analysis because it focuses more on the way that we decompose a system into parts, and in the way that those parts interact with one another, than with the behavior of the system in relation to some external object.

Specifically, Nagel says that when we describe a system as being goal directed, we analyze that system into a set of components. Consider the system that regulates the water content of the blood and keeps it at around 90 %. There are three components that are most directly relevant: the kidneys, the muscle, and the blood. When the water content of the blood starts to drop, the muscles release more water into the blood. When the water level gets too high, the kidneys extract more water from the blood. Each component can be described by a certain variable that can take different numeric values. One variable V_M represents the rate at which the muscles release water into the blood. Another variable V_K represents the rate at which the kidneys extract water. A third, the goal variable V_G, represents the water content of the blood. The important point is that when the goal variable departs from the goal-state (when V_G is greater than or less than 90 %) the other variables change in such a way as to bring the system back into the goal state. So, goal directedness has to do with the way the components of the system interact so as to maintain what he called the "goal-state."

(Incidentally, one might think that, formally speaking, Sommerhoff's analysis is just a special case of Nagel's analysis. Sommerhoff examines the relationship between a system variable V_S and an environmental variable V_E. He says that a system is goal directed with respect to goal G when the following counterfactual is true: if V_E were to change in certain ways, then V_S would also change in certain ways such that G is attained or maintained. In contrast, Nagel examines the components within a system, the V_i. He says that a system is goal directed under the following conditions: if V_G changes, the other V_i change in such a way as to maintain G. Perhaps Sommerhoff's analysis reduces to Nagel's analysis if we re-label Sommerhoff's "V_E" with Nagel's "V_G," that is, if we treat Sommerhoff's V_S and V_E as two components of a larger system. I think the only problem with the idea is that, as I noted above, Sommerhoff emphasizes a certain asymmetry between V_S and V_E. V_E changes more or less independently of V_S; V_S just adapts itself to V_E. But on Nagel's view, there is an inherent symmetry between the variables. They affect each other's behavior equally. For example, if the water content of the blood (V_G) changes, the rate at which kidneys extract water (V_K) also changes. But the converse is also true. If V_K changes, then so does V_G. At any rate, I think there is some interesting logical relationship between the two analyses but it is not entirely trivial to flesh it out.)

Let's return to the problem of overbreadth. Nagel recognized that his analysis of goal directedness suffered from the problem of overbreadth. Consider a pendulum at a state of rest. A gust of wind displaces it, and the pendulum swings back to rest. Is the pendulum goal directed with respect to the state of rest? After all, we can analyze the movement of the pendulum in terms of two variables, the force of displacement and the force of restoration, and we can describe the changing relationship between these variables in terms of the formal apparatus he recommends.

Nagel, borrowing from Sommerhoff's analysis, tries to avoid this problem by saying that the variables of a goal directed system must have the right sort of independence (Nagel 1953, 211). As Nagel (1977, 273) put it, there has to be an "orthogonality of variables." Specifically, at a given moment, the value of one controlling variable must not determine the value of any other controlling variable, *at that very moment*. At any point in time, each controlling variable must maintain some degree of freedom from the other controlling variables. Consider the system that regulates the water content of the blood. At any given moment, the rate at which the kidneys extract water is independent of the rate at which the muscles release water. We could test this claim by performing the right sort of intervention, for example, by artificially decreasing the rate at which the kidneys extract water. *Eventually* our intervention will affect the rate at which the muscles release water *but not at the very moment* we perform the intervention. By contrast, consider a pendulum swinging to a state of rest. The controlling variables include the force of displacement (away from the vertical position) and the force of restoration. Yet, at any given moment, these two forces must be equal and opposite, according to Newton's third law of motion. Thus, the pendulum's controlling variables lack the right sort of independence.

Other physical systems also lack the right sort of independence to qualify as goal directed. By Boyle's law, the pressure and volume of a given amount of gas, at a constant temperature, are inversely proportional. So, they are not epistemically independent. Nor are the voltage and current flow in an electrical circuit of constant resistance (by Ohm's law). So, the idea of epistemic independence seems to give us exactly what the other analyses lack. It gives us a plausible principle that we can use to differentiate real and fake goal directed systems.

Some commentators have attacked this maneuver by arguing that the very idea of epistemic independence is incoherent or that it fails to distinguish the different sorts of systems. Woodfield (1976, 67) criticizes it on the grounds that, in a genuinely goal directed process, the variables are never entirely independent of one another. Nissen (1980, 130) echoes this criticism. So long as the goal directed system is in functional condition, he argues, the variables will all be correlated with one another and hence derivable from one another. But I think this misses Nagel's point. Nagel's point was that, in a goal directed system, the variables are independently modifiable *at one and the same moment*. Again, go back to the system that regulates the water content of the blood. At any time, t, if I increase the water content of the blood at t, the behavior of the kidney will not change *at t*. It will change at some $t_1 > t$. But I cannot do this for the pressure and volume of a gas at constant temperature, or for the forces of displacement and restoration on a pendulum. The basic laws of nature do not allow one to modify independently the force of displacement and force of restoration of a pendulum. So it is not a goal directed system.

But problems remain. Presumably, even in a legitimate goal directed system, not *all* of the controlling variables will be independent. Humans have the goal of maintaining balance. The movement of fluid in the inner ear mediates this goal. The movements of the fluid are sensitive to the rotation of the head via basic physical laws (Newton's laws of motion) so at least some of the controlling variables that allow us to maintain balance are not independent. Moreover, at a certain level of description, even the atoms that control the movement of a rolling marble are somewhat independent of one another. Does that mean that, in a rolling marble, the controlling variables are epistemically independent? The problem here is that whether the set of controlling variables are independent of one another depends crucially on how we decompose the system. On some decompositions, they will be independent, and on others, they will not be. That suggests that whether or not a system is goal directed depends on how we choose to analyze it.

Interestingly, Nagel himself seemed to be perfectly happy with this outcome. In other words, he was willing to concede that whether a system is, or is not, goal directed partly depends on how we choose to analyze it (Nagel 1977, 275). According to this way of thinking, the notion of goal directedness is part of an *analytic strategy* for coming to grips with complicated systems. It is a useful construct that helps us to make sense of systems that are otherwise intractably complex, but it is not a mind-independent fact of nature. Imagine trying to describe and predict the complex movements of the frog's head, retina, and tongue, without describing the frog as having the goal of catching flies. It can scarcely be done.

The biologist McShea (2012) recently defended the value of goal directedness for studying complex systems. He himself develops a viewpoint that is similar in some ways to Braithwaite's in that it emphasizes the role of the environmental variables in bringing about the adaptive behavior. His view is that all of the traditional examples of goal directedness, such as homing torpedoes, fly-catching frogs, and even human intentions, exemplify the very same physical structure. In each case, there is a larger entity that constrains the behavior of a smaller entity within it, without determining it precisely. He uses an example of a bacterium moving up a concentration gradient. The larger system is the concentration gradient, and the smaller system is the bacterium. The concentration gradient influences the behavior of the bacterium in such a way that it appears goal directed. He believes that a similar dynamic plays out in the context of evolution by natural selection and even human intentional behavior.

Now, McShea does not purport to offer a conceptual analysis of "goal directedness." Rather, he is simply trying to identify, empirically, a common structure that underlies most cases of goal-directedness, but which also underlies some behaviors that do not strike us as goal directed (such as a swinging pendulum). The important part, for my purposes, is that he is happy to concede that, if one were forced to distinguish precisely between systems that are goal directed and those that are not, one would have to do so on pragmatic grounds. As he puts it, in discussing the problem of overbreadth, "the present suggestion that goal directedness is a function of the perceived complexity of the system is likewise relative to our knowledge" (682). Trestman (2012, 215) also suggests that what counts as a goal directed system is relative to our perspective.

I have to admit that I find this pragmatic and, if you will, anti-realist line unsatisfying. The reason is simple. It seems to me that whether or not frog has the goal of catching flies does not depend on how much knowledge we happen to have of it, but is intrinsic to it. Perhaps a few hundred years from now, the frog's ability to catch flies will seem as simple to scientists, then, as swinging pendulums seem to us, now, and scientists will find no use for analyzing the system as goal directed. But I take it that frogs will still have goals. This is not just because the frog has conscious intentions, which I believe it does. It is because whatever (behavioral, psychological, or mechanistic) hallmarks we are picking out *now* that explain our judgment that the frog is goal directed will still be true of it hundreds of years from now when its behavior can be entirely explained at the level of sub-atomic physics and when the language of goals and purposes fails to yield any novel predictive benefits. But I appreciate that we are wandering into more general issues in the philosophy of science pertaining to realism and antirealism that cannot be resolved here.

References

Adams, F. R. (1979). A goal-state theory of function attributions. *Canadian Journal of Philosophy, 9*, 492–518.

Boorse, C. (1976). Wright on functions. *Philosophical Review, 85*, 70–86.

Braithwaite, R. B. (1953). *Scientific explanation*. Cambridge: Cambridge University Press.

Canfield, J. (1966). Introduction. In J. Canfield (Ed.), *Purpose in nature* (pp. 1–7). Englewood Cliffs, NJ: Prentice-Hall.

Ehring, D. (1984). The system-property theory of goal directed processes. *Philosophy of the Social Sciences, 14*, 497–504.

Faber, R. J. (1986). *Clockwork Garden: On the mechanistic reduction of living things*. Amherst: University of Massachusetts Press.

Gilbert, S. F., & Sarkar, S. (2000). Embracing complexity: Organicism for the 21st century. *Developmental Dynamics, 219*, 1–9.

Kandel, E. R., et al. (2013). *Principles of neural science* (5th ed.). New York: McGraw Hill.

McShea, D. W. (2012). Upper-directed systems: A new approach to teleology in biology. *Biology and Philosophy, 27*, 663–684.

McShea, D. W. (2013). Machine wanting. *Studies in History and Philosophy of Biology and Biomedical Sciences, 44*, 679–687.

Nagel, E. (1953). Teleological explanation and teleological systems. In S. Ratner (Ed.), *Vision and action* (pp. 537–558). New Brunswick, NJ: Rutgers University Press.

Nagel, E. (1961). *The structure of science*. New York: Harcourt, Brace and World.

Nagel, E. (1977). Teleology revisited: Goal directed processes in biology and functional explanation in biology. *Journal of Philosophy, 74*, 261–301.

Nissen, L. (1980). Nagel's self-regulation analysis of teleology. *Philosophical Forum, 12*, 128–138.

Nissen, L. (1993). Four ways of eliminating mind from teleology. *Studies in History and Philosophy of Science A, 24*, 27–48.

Piccinini, G. (2015). *Physical computation: A mechanistic account*. Oxford: Oxford University Press.

Rignano, E. (1931). The concept of purpose in biology. *Mind, 40*, 335–340.

Rosenblueth, A., Wiener, N., & Bigelow, J. (1943). Behavior, purpose and teleology. *Philosophy of Science, 10*, 18–24.

Russell, E. S. (1945). *The directiveness of organic activities*. Cambridge: Cambridge University Press.

Schaffner, K. (1993). *Discovery and explanation in biology and medicine*. Chicago: University of Chicago Press.

Scheffler, I. (1959). Thoughts on teleology. *British Journal for the Philosophy of Science, 9*, 265–284.

Sommerhoff, G. (1950). *Analytical biology*. London: Oxford University Press.

Sommerhoff, G. (1969). The abstract characteristics of living systems. In F. E. Emery (Ed.), *Systems thinking* (pp. 147–202). Middlesex: Penguin.

Sommerhoff, G. (1974). *The logic of the living brain*. New York: Wiley.

Taylor, R. (1950). Comments on a mechanistic conception of purposefulness. *Philosophy of Science, 17*, 310–317.

Trestman, M. A. (2012). Implicit and explicit goal directedness. *Erkenntnis, 77*, 207–236.

von Bertalanffy, L. (1950). An outline of general systems theory. *British Journal for the Philosophy of Science, 1*, 134–165.

Wiener, N. (1948). *Cybernetics*. Cambridge, MA: MIT Press.

Wimsatt, W. C. (1971). Some problems with the concept of feedback. *Boston Studies in the Philosophy of Science, 8*, 241–256.

Wimsatt, W. C. (1972). Teleology and the logical structure of function statements. *Studies in the History and Philosophy of Science, 3*, 1–80.

Woodfield, A. (1976). *Teleology*. Cambridge: Cambridge University Press.

Chapter 3
Function and Selection

Abstract This chapter focuses on the selected effects theory of function. According to this view, a function of a trait is whatever it was selected for by natural selection or some natural process of selection. I show how the theory plausibly accounts for the explanatory and normative aspects of function. First, if a function of a trait is whatever it was selected for by natural selection, then when one attributes a function to a trait one provides a causal explanation for why the trait currently exists. Second, since the function of a trait is determined by its history rather than current performance, it is easy to see how a trait can have a function that it cannot perform ("dysfunction"). I sketch the somewhat complex historical background of this theory. The theory was actually developed by biologists throughout the twentieth century, and in the 1970s philosophers began to explore it systematically. I survey the major criticisms of the theory and show why they are not compelling. Critics say that it does not really account for the explanatory and normative features of function; that it is inconsistent with the way biologists actually use the term; that there are (real or imaginary) counterexamples; and that it is committed to adaptationism. I close by presenting a new version of the theory, the generalized selected effects theory, which shows how brain structures (such as synapses) can acquire new functions during an individual's lifetime through a process that is analogous in some ways to natural selection itself.

Keywords Selected effects theory · Etiological theory · Teleofunctions · Backwards-looking functions · Neural selection · Synapse selection

3.1 What (and How) Do Functions Explain?

This chapter is about the selected effects theory of function. The selected effects theory begins with an observation about how some biologists use the term "function." This observation creates a puzzle. The selected effects theory purports to solve that puzzle. This is not the only sense of "function" that biologists ever use. Many prominent selected effects theorists are also function pluralists (see Sect. 5.3).

© The Author(s) 2016
J. Garson, *A Critical Overview of Biological Functions*,
Philosophy of Science, DOI 10.1007/978-3-319-32020-5_3

However, sometimes biologists do use the term "function" in this puzzling way. The fact that they use it this way *sometimes*, in and of itself, raises significant philosophical problems.

An observation that motivates the selected effects theory is that, at least in some contexts, functions purport to be explanatory. Sometimes, when I state the function of an item, I purport to explain why that item exists, and I do so by pointing to something the item does. On the surface, this looks like an explanation in the straightforward causal sense of the term. For example, when I say that the function of the nose is to help us breathe, rather than hold up glasses, I am trying to provide a causal explanation for why people have noses. I am trying to answer a "why-is-it-there" question. People have noses because noses help them breathe.

Some philosophers, such as Wouters (2013, 480), are skeptical that biologists *ever* use the term function in this way. This leads him to suggest that the selected effects theory (which I'll explain shortly) is based on a problem that doesn't really exist. A clear example might help to dispel that worry. The American biologist Tim Caro and his colleagues recently published a study on the function of zebra stripes, entitled simply, "The Function of Zebra Stripes" (Caro et al. 2014). In it, he argues that zebra stripes have the function of deterring biting flies (particularly Tsetse flies which carry a virus responsible for sleeping sickness). According to this research team, Tsetse flies, along with some other species of biting fly, are averse to landing on striped objects. Nobody entirely knows why this is the case.

It is clear from the text that for Caro, to solve the riddle about the function of stripes is just to discover why zebras have stripes, in some causal, how-they-got-to-be-there sense [what he calls their "evolutionary drivers" (3)]. For example, he describes five different "functional hypotheses" for zebra stripes, and he also refers to these as "factors proposed for driving the evolution of zebras' extraordinary coat coloration" (2). That suggests that, for Caro, when we state something's function we are giving a causal explanation for why it is there. Caro and his colleagues were also very careful to gather *historical* data about Tsetse fly and predator distributions. This would be strange if they only wanted to show how stripes happen to benefit zebras currently. But it is not strange on the supposition that they wanted to explain, causally, why zebras have stripes. My point is not to assess the empirical merits of Caro's theory, which has been criticized by others (Larison et al. 2015). My point is that Caro's article nicely illustrates a genuine strain of biological usage of the term "function," and we should take this usage seriously.

Moreover, the journalists who reported the discovery often used the phrase "the function of stripes" interchangeably with phrases like "why stripes evolved," or "why zebras have them," or "the reason for stripes."[1] This shows that this sense of function is deeply embedded in popular as well as biological usage.

---

[1]E.g., http://www.telegraph.co.uk/news/earth/wildlife/10737443/Why-do-Zebras-have-stripes-Scientists-claim-to-have-the-answer.html, accessed February 16, 2015. Also see http://www.theguardian.com/science/animal-magic/2014/apr/02/why-do-zebras-have-stripes-scientists-have-the-answer, accessed February 16, 2015.

So much for the observation. Now for the problem. How can an effect of the trait explain its own existence? Logically, in order for the zebra's stripes to deter biting flies, zebras must already have stripes. So the fact that stripes deter flies, it would seem, cannot possibly explain why zebras have them. This is the infamous problem of backwards causation.

One way of solving this problem is to appeal to natural selection. Zebras have stripes because (let us suppose), in the past, there was a population of ancestors of modern zebras, some of which had stripes and some of which did not. Those that had stripes had a fitness advantage over those with some other pattern of coat coloration (because they were better at deterring biting flies), and hence that trait persisted in the population. That is why zebras generally have stripes today, rather than some alternate pattern of coloration (e.g., non-striped species of the same genus, such as horses).

Let's suppose that selectionist story is correct. That has a remarkable consequence. When we appeal to natural selection, we cite an effect of stripes as part of an explanation for why zebras have stripes today. The problem of backwards causation dissolves. We can legitimately appeal to something that stripes *do* in order to explain why zebras have them. To be specific, we appeal to something that stripes did in the past to explain why zebras have them today.

The reader may suspect that there is some sort of elementary confusion involved here. The problem of backwards causation stems from the observation that sometimes scientists explain why a trait is there (why the members of a species have a trait today) in terms of something the trait *does*. The selected effects theorist says that, if selection is involved, then we can explain why a trait is there in terms of something it does (e.g., stripes are there because stripes deter flies). But technically, when we give a selectionist account for why zebras have stripes, we explain why stripes are there in terms of something the stripes *did in the past*, rather than something it does *now*. So, the critic argues, there is some temporal smudging going on that lends false plausibility to the selected effects theory.

On reflection, however, there is no trickery or deception going on. When I say, "stripes are there because of something they do," I do not use the term "do" in the present tense (i.e., something they are doing at this very moment). I mean to use the term "do" more generally, in a tenseless way. By analogy, when I say, "hawks eat fish," I do not mean that some hawks are eating some fish at this very moment. So long as enough hawks ate enough fish recently, then it is true to say that "hawks eat fish," even if they are not doing so right now.

Any reader who briefly consults the three desiderata that I identified in Sect. 1.2 (the function-accident distinction, the explanatory feature of functions, and the normativity of functions) will quickly see that the selected effects theory, at least on the surface, satisfies those desiderata in a very natural way. (In the remainder of this chapter I will elaborate some of these key points and defend them against some criticisms.) First, it distinguishes between function and accident. The function of the zebra's stripes is to deter biting flies and not to provide entertainment for spectators on safari because only the former explains (via natural selection) why zebras have stripes.

Second, it accounts for the explanatory dimension of functions. As I discussed in Sect. 1.2, one traditional desideratum for a theory of function is that it makes sense of how functions can be explanatory. If function statements cite the effect that a trait was selected for, then they constitute causal explanations for the existence of that trait, and hence they satisfy this explanatory desideratum (bearing in mind that there may be other, non-causal senses of explanation). Sometimes the selected effects theory is called an "etiological" theory for this reason, that is, because functions have to do with the causal history of the trait.

Third, the selected effects theory accounts for the normative dimension of functions. To say that functions are normative means that it is logically possible for a trait token to have a function it cannot perform. But according to the selected effects theory, the function of a trait is determined entirely by the trait's history, rather than its current abilities. So, it is obvious how something can have a function it cannot perform. These considerations provide a strong case that the selected effects theory captures one important strand of biological usage.

Before I go on to the next section, I should elaborate the idea of a teleological explanation, as I promised to do earlier (Sect. 2.1). Sometimes philosophers of biology refer to the selected effects theory as a teleological account of function. I think that this label is technically correct, though nothing in my argument relies on this terminological point. As I understand it, *a teleological explanation is one that purports to explain the existence of an entity (such as an organism, trait, or behavior), in terms of some effect the entity brings about.* So, when we say, "the reason zebras have stripes is because ..." and we go on to describe something good that stripes do, we are offering a teleological explanation for stripes. Stripes are there because they help zebras in this special way. In Aristotle's famous example of a teleological explanation, teeth are there because they are good at chewing.

If this is the right way of thinking about teleological explanation, then selectionist explanations for a trait are teleological explanations. To recap, suppose we explain why zebras have stripes today by saying that, in the past, stripes were selected for deterring flies. Then we are citing an effect that stripes produce (specifically, one they produced in the past) as part of an explanation for why zebras have them. But this is just a teleological explanation. If function statements are selectionist explanations, and selectionist explanations are teleological explanations, then function statements are teleological explanations, too. (Ayala 1970; Wimsatt 1972 make this point. I will discuss their view of teleological explanation in the next section.)

I realize that some people would balk at my liberal understanding of what a teleological explanation is. They would say that a teleological explanation, in its true sense, requires backwards causation. It requires some mysterious causal influence from the future reaching back into the past. Since there is no backwards causation, then, in their view, all teleological explanations are mistaken. If we stick with that restrictive sense of teleological explanation, I agree that all teleological explanations are mistaken. If we accept my broader sense of teleological explanation, I think there are some correct teleological explanations. At any rate, I do not want to get hung up on the right way to define "teleological explanation." Rather, I simply wish to clarify the broad sense of "teleological explanation" that proponents of the selected effects theory sometimes use.

3.2 Historical Development of the Selected Effects Theory

The selected effects theory was developed throughout the course of the twentieth century by philosophically-minded biologists who reflected on the notion of function and the corresponding notion of purpose. The historian Lennox (1993) argues that Darwin himself deliberately and self-consciously used terms such as "end," "purpose," and even "final cause," synonymously with the effect that a trait was selected for. Strikingly, Darwin's friend, the botanist Asa Gray, once wrote of "Darwin's great service to natural science in bringing back to it Teleology." What is even more striking is that Darwin responded quite approvingly: "What you say about Teleology pleases me especially and I do not think anyone else has ever noticed the point" (cited in Lennox and Kampourakis 2013, 437). The oft-stated claim that Darwin removed teleology from biology is not quite correct. At least in a broad sense of teleology, he vindicated teleological explanations by showing how they could be grounded in selection processes.

Many other biologists made similar points. The British neuroscientist Charles Sherrington, in his 1906 book, *The Integrative Action of the Nervous System*, echoes Gray's observation:

That a reflex action should exhibit purpose is no longer considered evidence that a psychical process attaches to it; let alone that it represents any dictate of "choice" or "will." In light of the Darwinian theory every reflex *must* be purposive. We here trench upon a kind of teleology...The purpose of a reflex seems as legitimate and urgent an object for natural inquiry as the purpose of the colouring of an insect or a blossom (235–6).

The ethologist Konrad Lorenz makes a similar remark in his 1963 book, *On Aggression*:

If we ask "What does a cat have sharp, curved claws for?" and answer simply "To catch mice with," this does not imply a profession of any mythical teleology, but the plain statement that catching mice is the function whose survival value, by the process of natural selection, has bred cats with this particular form of claw. Unless selection is at work, the question "What for?" cannot receive an answer with any real meaning (Lorenz 1966 [1963], 13–4; cited in Griffiths 1993, 412).

The evolutionary biologist George Williams also emphasizes this point: "The designation of something as the *means* or *mechanism* for a certain *goal* or *function* or *purpose* will imply that the machinery involved was fashioned by selection for the goal attributed to it" (Williams 1966, 9). The biologists Gould and Vrba (1982, 6) also endorse Williams' use of "function." Similarly, the biologist Maynard Smith (1990) tells us that:

...when a biologist says 'The heart beats in order to pump blood round the body', this is a short-hand for 'Those animals which, in the past, had hearts that were efficient pumps survived, because oxygen reached their tissues, whereas animals whose hearts were less efficient as pumps died. Since offspring resemble their parents, this resulted in the fact that present-day animals have hearts that are efficient pumps' (66).

He also says that, "by 'function,' I mean those consequences of a structure (or behaviour) that, through their effects on survival and reproduction, caused the evolution of that structure" (67).

Most recently, the geneticist Doolittle (2013) wrote:

Most philosophers of biology, and likely, most practicing biologists when pressed, would endorse some form of the selected effect (SE) definition of function...the functions of a trait or feature are all and only those effects of its presence for which it was under positive natural selection in the (recent) past and for which it is under (at least) purifying selection now. They are why the trait or feature is there today and possibly why it was originally formed (5296).

These citations, taken collectively, show that the selected effects theory represents a certain tradition of thinking about function and teleology among biologists that stretches back for at least a century. This is not a philosophers' invention. In the 1970s, however, philosophers of science began to explore systematically the relationship between function statements and selectionist explanations. They were taking their cue from what biologists were already doing, rather than trying to impose some philosophical agenda on science.

One of the first of these philosophers was Francisco Ayala, who is also a practicing biologist. In two papers, he explicitly notes that selectionist explanations are teleological explanations (Ayala 1968, 1970). As he points out, a teleological explanation is one in which "the presence of an object or a process in a system is explained by exhibiting its connection with a specific state or property of the system to whose existence or maintenance the object or process contributes" (Ayala 1970, 8). But, "the adaptations of organisms ... are explained teleologically in that their existence is accounted for in terms of their contribution to the reproductive fitness of the organism" (9). Though he notes the connection between natural selection and teleology, he does not *define* "function" explicitly in terms of natural selection.

The philosopher of biology William Wimsatt also explored the intimate connection between natural selection and functions. His view, set forth in 1972 and developed in several papers, appears to be a version of the goal-supporting theory of functions, which I described in the previous chapter (Wimsatt 1972, 2002, 2013). But he notes that the concepts of purpose and teleology must both be "at least partially explicated by bringing in a third feature—that of a selection process" (12). He then goes on to say, "the operation of selection processes is not only *not* special to biology, but appears to be at the core of teleology and purposeful activity wherever they occur" (13). Wimsatt uses the notion of selection very broadly, to refer to all of those processes that involve what he called "blind variation" and "selective retention." These include learning by trial and error, natural selection, and even the sort of mental or "virtual" trial and error that takes place when we reason through various courses of action before selecting one.

Strangely, Wimsatt refused to *define* "function" in terms of selection (even in this generalized sense of selection). This is because he was working under the assumption that an explication of function (or purpose) should be a conceptual analysis of ordinary linguistic usage. But ordinary people do not define "function" in terms of selection. First, he reasoned, being the product of a selection process is not *necessary* for having a function (or purpose). After all, if God is real, then God would not have to go through any sort of mental trial-and-error process before making decisions. God would just know what to do. But God's creations would still have functions (15). Note, however, that this objection is only compelling if we think that one and the same theory must account for both artifact functions and biological functions. It seems to me that if God created everything in its present state, then nothing would actually have a biological function. Rather, entities would only have artifact functions. Wimsatt's conclusion, which was supposed to be a *reductio ad absurdum*, seems plausible to me if we restrict attention to biological functions.

Second, Wimsatt reasoned that selection is not *sufficient* for having a function. Stars undergo differential survival (that is, they survive at different rates). This could plausibly be called a type of "selective retention." But stars do not have functions. (I will come back to this interesting concern in Sect. 3.4 where I develop a view of functions very much like the one Wimsatt entertains here.) I am not entirely convinced by this counterexample, because I do not think that stars form populations in the sense required for functions. I will explain this point in the next section.

Larry Wright developed a view of function and recognized the connection between function and natural selection. As noted in the last section, Wright did not *define* "function" in terms of natural selection. This is because he, like Wimsatt, was working under the idea of unification, that is, that an analysis of function should be neutral between biological and artifact functions: "functional ascriptions of either sort have a profoundly similar ring" (143). He didn't think that artifacts needed to go through anything like a selection process in order to acquire new functions. The hammer's claw has the function of pulling out nails. That is not because, long ago, people made some hammers without claws and some hammers with claws, and the hammers with claws sold more quickly than those without it, and so on. So "function" cannot be defined in terms of selection processes. As I noted in Sect. 1.2, I agree that such a unifying theory of function would be a virtue, but I do not think that it represents some sort of adequacy condition on an acceptable theory of function.

Instead, Wright defined function in terms of a very general process that he called a "consequence-etiology." As he put it (changing the variables for consistency):

The function of T is F *means*
  (a) T is there because it does F,
  (b) F is a consequence (or result) of Ts being there (161).

In the biological case, natural selection is the mechanism, as it were, that grounds the truth of function ascriptions. In other words, in the biological case, the first clause, "*T* is there because it does *F*," is true when *T* was selected for *F* (159).

As the reader will notice, Wright has a second clause in his definition, namely, that *F* results from *T*s being there. I think this clause is unnecessary and tends to obscure the point of the analysis. Why did he add it? It is certainly *not* the case that he added it because he thought, along the lines of the fitness-contribution theory, that traits must currently be able to perform their functions. Rather, if I understand him correctly, he wanted to exclude cases where *F* represents something like a *requirement* for the trait's being there, rather than an effect of that trait. For example, a reason we have oxygen in our blood is because oxygen binds to hemoglobin. This is a requirement for oxygen's being in our blood. So it is true, in some sense, to say that oxygen "is there" (in the blood) because it binds to hemoglobin. But that is not its function. Its function is to energize cells. Perhaps a slightly different example could illustrate the point better. Brain activity utilizes about 20 % of our daily caloric intake. I suppose this is a requirement for the brain to exist and function well. But it sounds strange to say that a function of the brain is to burn calories. It seems more natural to say that a function of the brain is to process sensory information efficiently. The fact that it burns so many calories is something like a requirement for its existence, rather than a function. This is what Wright is getting at in his condition (b).

Both of Wright's conditions, I believe, can easily be collapsed into one:

The function of *T* is *F* if and only if *T* is there because *T* results in *F*.

But read this way, it becomes clear that Wright's second condition merely clarifies the sense of "does" in the first condition. It does not add any substantial new content.

The most influential objection to Wright's analysis came from a paper by Boorse (1976). As I noted in the last chapter, Boorse developed his own goal-contribution theory of function. Although Boorse outlined several problems, the most serious was a series of clever counterexamples that exposed the insufficiency of Wright's conditions. Suppose that a hose in a laboratory springs a leak and emits a noxious chemical, and any scientist that attempts to seal it gets knocked unconscious by the chemical it emits. The leak in the hose contributes to its own persistence by knocking out anyone that comes close enough to fix it (Ibid., 72). That is, the leak is there because it results in knocking out scientists. But that is not its function.

Similar counterexamples abound. Obesity contributes to a sedentary lifestyle, which reinforces obesity. So one can explain a person's current obesity in terms of one of the consequences his or her obesity produced in the past that contributes to its own persistence (Ibid., 75–6). Yet, like the hose example, obesity does not have the function of contributing to a sedentary lifestyle. Bedau (1992, 786) used an example of a stick floating down a stream that brushes against a rock and gets pinned there by the backwash it creates to make the same point. Clearly, trivial examples from the natural world as well as from the realm of artifacts can be multiplied indefinitely.

One way to exclude these sorts of counterexamples is just to *define* "function" in terms of selection. This is the conclusion that both Neander (1983) and Millikan (1984) independently arrived at. Hence, they gave us the first canonical statements of the selected effects theory of function. This move avoids the Boorse-type counterexamples. After all, obesity was not *selected for* contributing to a sedentary lifestyle. It is not as if there was a population of people, some of which were obese and some of which were not, and the ones that were obese had more sedentary lifestyles, which somehow contributed to their fitness and hence caused some genes for obesity to persist. Brandon (1981), like Ayala (1970), came to a very similar conclusion in his discussion about teleology generally, though he was not specifically concerned with defining "function." In his view, biologists can legitimately answer "what-for" questions by citing selection history.

Though Neander and Millikan agree on the basic contours of a theory of function, they disagree about specifics. First, as noted in Sect. 1.3, Neander presents her theory as a conceptual analysis of modern biological usage, while Millikan presents hers as a theoretical definition. Second, and more importantly, Neander's view focuses on selection in the ordinary evolutionary sense of the term, that is, natural selection taking place over evolutionary time scales, mainly by altering the genetic composition of the population: "It is the/a proper function of an item (X) of an organism (O) to do that which items of X's type did to contribute to the inclusive fitness of O's ancestors, and which caused the genotype, of which X is the phenotypic expression, to be selected by natural selection" (1991, 174). Millikan accepts a much broader construal of selection, as I will shortly describe.

One consequence of Neander's definition is that she does not think that functions are the same in biology and in artifacts. She recognizes that artifacts typically do not arise by any process that resembles natural selection. Over the decades, artifacts such as televisions are modified in their basic design and these modifications are often cumulative. But the disanalogies are glaring. For example, the very first instance of an artifact such as a vacuum tube or an arrowhead possessed a function, namely, whatever function its creator gave it. But the very first instance of, say, zebra stripes had no function at all, because natural selection had not yet taken place.

Neander (1991, 175) acknowledges that there may be some loose, vague notion of "selection" that would apply to organisms and artifacts, but she is not very interested in explicating that notion. Griffiths (1993) develops the analogy in more detail. Over the last few decades, some debate has arisen regarding whether, and the extent to which, we can use selection processes to make sense of artifact functions, and more generally, whether biological functions and artifact functions are the same sort of thing (Preston 1998, 2003; Millikan 1999; Vermaas and Houkes 2003; Lewens 2004; Piccinini 2015, Chap. 6; Maley and Piccinini, Submitted for publication; see Krohs and Kroes 2009 for an anthology devoted to the topic of biological and artifact functions). As I noted at the outset (Sect. 1.4), I will not pursue this interesting literature here.

In contrast, Millikan (1984, 28) wants the notion of selection to be understood quite broadly, closer to the liberal sense in which Wimsatt (1972) used the term. In her view, functions are restricted to certain sorts of populations, namely, groups of items that she refers to as "reproductively established families." Two entities belong to the same reproductively established family if one is a copy of the other. Two instances of the same gene sequence, in, say, the mother and the child, are part of the same reproductively established family because the one is a copy of the other (produced by genetic copying mechanisms). But two instances of a gesture, say, a handshake, can belong to the same reproductively established family if the one is a copy of the other (here, produced by mechanisms of cultural transmission). For example, if my older son learned to shake hands by imitating me, then a token handshake performed by me, and one performed by him, are members of the same reproductively established family. These handshakes also form something like a lineage.

With this conceptual background in mind, when we say that a token *t* has the "direct proper function" *F*, what we are saying is that, in the past, tokens of that type did *F*, and that one of the reasons that *t* exists today is because its ancestors (that is, earlier copies of the trait) performed *F better than* other tokens (Millikan 1984, 28). I believe that her original formulation is fairly dense, but that this is a fair rendering of it.

Some formulations of her view have been ambiguous. Specifically, in some contexts, she is not entirely clear that something like selection is required, and her view collapses into what Buller (1998) calls the "weak etiological theory" (see Sect. 6.1). For example, Millikan (1989a, 288) is not as clear in her formulation of the definition that selection must be involved. All that is required is that *t* exists in part because *t*'s ancestors did *F*. Millikan (1989b, 199), however, clarifies that selection *must* be present: "a trait's function is what it actually did—did most recently—that accounts for its current presence in the population, as over against *historical* alternative traits no longer present." Also, Millikan (1993, 35–6) emphasizes that in order to have a function, it is not enough that an item *somehow* contribute to its own reproduction but that it be selected for: "simply cycling through reproductions … doesn't intuitively seem to be having a function in any reasonable sense…. Only if an item or trait has been *selected* for reproduction, *as over against other traits*, *because* it sometimes has a certain effect does that effect count as a function."

There is one more feature of Millikan's view that deserves mention here, which is the notion of a "derived" proper function. Suppose a species excels at changing its skin pattern for camouflage. The cuttlefish (which is actually a kind of mollusk) provides an excellent example of camouflage. Each chromatophore (pigment-filled cell) is surrounded by muscle cells and it can expand or contract independently of the other chromatophores. This allows it to generate a vast diversity of new patterns.

Now suppose, on a given occasion, a cuttlefish develops a pattern of camouflage, *c*, that is entirely novel, that is, that no cuttlefish in the history of the earth has ever exhibited. In this case we cannot say that *c*, described in its specific detail, has what she calls a "direct proper function." Since *c* is unprecedented it cannot have been selected for. Rather, Millikan would say, it has a *derived* proper function. It has the derived proper function of camouflage because the mechanisms that produce it (the brain mechanisms that control the muscle) have the *direct* proper function of camouflage, and they typically carry out this function *by means of* producing novel patterns such as *c* (Millikan 1984, 41–2, 1989a, 288, 2002, 125–131). In this way she is able to avoid the apparently absurd implication that the novel pattern of coloration, described in its specificity, has no function.

There are two other developments that deserve mention here. One problem for the selected effects theory, as I have outlined it, is that it does not account for the possibility of vestiges. A vestige is a trait that once had a function but lost it. The human appendix is probably a vestige that once had the function of harboring healthy gut bacteria. But if the appendix was selected for, in the distant evolutionary past, because it harbored gut bacteria, then one reason humans have appendices (now) is because they were selected for harboring bacteria. According to the selected effects theory, that would constitute their function. So what went wrong?

Griffiths (1992, 1993) and Godfrey-Smith (1994) both try to solve the problem by arguing that, in order to have a function, the trait must have undergone selection in the *recent* evolutionary past. That seems sensible enough, but how recent is "recent?" A year? Ten thousand years? Ten million? Any precise cutoff seems arbitrary.

Griffiths attempts to quantify this unit of time, which he calls an "evolutionarily significant time period," more precisely. An evolutionarily significant time period for a trait *T* is a time period such that, given the mutation rate at the genetic loci controlling *T*, we would have expected enough mutations to arise that *T* would have atrophied, were it not contributing to fitness. In other words, we simply ask: under the assumption that *T* does *not* contribute to fitness, how much time should we expect to pass before *T* atrophies? That unit of time is called an "evolutionarily significant time period" with respect to some function *F*. We can then say that *T* is a vestige (*simpliciter*) if it has not contributed to fitness during the last evolutionarily significant time period (Griffiths 1992, 128; 1993, 417). Godfrey-Smith (1994) also emphasizes recent history, but does not try to specify what counts as recent in any precise way.

Schwartz (1999, 2002) sees a problem for the recent history view. I will call it the "no-variation" problem. He claims that, intuitively, there could be a trait *T* that underwent selection for some activity *F* in the distant evolutionary past, and now has *F* as its function, but has not undergone selection for *F* recently. One way this might happen is if *T* went to fixation in the distant evolutionary past, and *T* continues to benefit the organism, but, by sheer accident, no genetic variation happened to occur between the time of its fixation and the present day. He thinks the mere possibility of such a trait undermines the recent history theory of functions.

One might wonder how biologically plausible this sort of no-variation case is. Why would it be the case that a trait varied a long time ago (enough for selection to occur) and then stopped varying? Moreover, keep in mind that, for the selected effects theory, stabilizing selection is good enough for maintaining a function. Take the mechanisms underlying neural tube formation in humans. Deviations from the normal process of neural tube formation are almost always fatal. This is an example of stabilizing selection. It seems overwhelmingly probable that, if some trait T was selected for long ago, and it still benefits us today, then some stabilizing selection for T would have taken place between then and now. Kraemer (2014), however, tries develop a plausible example of a no-variation case.

Suppose we take these no-variation cases seriously, even as a theoretical possibility. Schwartz' solution is to emphasize the difference between a trait's making a *contribution to fitness*, and a trait's being *selected for*. (I will return to this distinction in Sect. 6.1 because it is central for the weak etiological theory.) A trait makes a *contribution to fitness* (in an absolute sense) when it contributes to survival or reproduction. In this absolute sense, a trait can contribute to fitness even when there is no variation for the trait. The human kneecap protects the knee joint, and it would still do that even if there were no variation for kneecaps. A trait undergoes *selection* when it contributes to *differential* survival and reproduction. In order for a trait to undergo selection there must be variation.

With this distinction in mind, Schwartz (1999, S219) amends the recent history view along the following lines. Trait T has function F if and only if:

 (i) T was *selected for* F at some time in the past (even distantly), and;
 (ii) T recently *contributed to fitness* by doing F.

The nice part of his definition is that it avoids the no-variation cases. A trait need not have undergone variation in the recent past in order to have a function. It just needs to have contributed to fitness.

Another way of responding to Schwartz, however, is simply to bite the bullet and assert that, if a trait was selected for some activity F a long time ago, but has not undergone recent selection for F, then F is no longer its function, even if having F continues to benefit the organism. F would, today, have the status of being a lucky accident rather than a function. Why couldn't we just call it a lucky accident? After all, the selected effects theorist is willing to admit that if a trait goes to fixation by genetic drift, and then, after going to fixation, happens to confer some benefit onto an organism, it would count as a lucky accident and not a function. So why couldn't we say the same thing about a trait that was selected for long ago, and that stopped varying, but that continues to benefit the organism?

A second development stems from the following consideration. There are certain systems that do things that cause them to get reproduced over others, but we hesitate to attribute functions to them. Consider cancer cells. Cancer cells subvert normal regulatory mechanisms and manage to replicate themselves more quickly than non-cancerous cells. This is a kind of selection process (sometimes described as "somatic selection"). But we typically do not say that cancer cells have the function of subverting cell-regulatory mechanisms. Why not?

Another example is meiotic drive (or segregation distortion). This takes place when one chromosome manages to subvert the normal process of meiosis and produce more copies of itself, which increases its chances of getting passed on to offspring (Crow 1979; see Godfrey-Smith 1994 for discussion). Segregation distortion genes can be harmful to the organism. For example, they can cause sterility in flies. We typically do not attribute functions to segregation distorting genes.

A related example comes from the study of transposable elements in genomes (Doolittle 1989; Elliott et al. 2014). These are segments of DNA that manage to duplicate and re-insert themselves in other parts of the genome in a "selfish" way. We usually do not attribute functions to them. Why not?

One way to block the attribution of functions to systems such as cancer cells, or segregation distorter genes, is to insist that, in order for something to have a function, it must contribute to the fitness of a larger, containing system (Godfrey-Smith 1994). In other words, it cannot just do something that causes its own differential reproduction in some selfish way. It has to contribute to its own reproduction *by* contributing to the fitness of a larger system.

Price (2001) develops a slightly different approach. Her idea is that, in order to have a function, a trait must contribute to the operation of a *second* system, which, in turn, contributes to the reproduction of the trait in question. The difference between the two views is a bit subtle. On Godfrey-Smith's view, a trait like the heart has a function because of the way it contributed to the fitness of a larger system (e.g., the organism). On Price's view, a trait like the heart has a function because of the way it contributed to *other* systems (such as the brain, lungs, and liver) that thereby contributed to the first. Whichever direction we go, it seems that the selected effects theorist should hold that a trait comes to possess a function by virtue of its contribution to the working of another system (or a more inclusive system).

3.3   Criticisms of the Theory

In this section I will lay out four major criticisms of the selected effects theory. I will also explain why I do not find them compelling. The first, and most important, is that the selected effects theory cannot, in fact, satisfy the desiderata it purports to. Specifically, according to this group of arguments, it cannot make sense of the explanatory and normative dimensions of function (Cummins 1975, 2002; Nagel 1977; Cummins and Roth 2010; Davies 2000, 2001, 2009). The second is that the theory runs counter to actual biological usage (Amundson and Lauder 1994, 446; Godfrey-Smith 1994, 351; Walsh 1996, 558; Schlosser 1998, 304; Wouters 2003, 658; Sarkar 2005, 17; Griffiths 2006, 3). The third is that the theory is susceptible to various real or imaginary counterexamples. The fourth is that, if everyone accepted it, it would impose serious epistemic burdens on biologists. In other words, if "function" means roughly the same thing as "adaptation," then it would be very hard to figure out what a trait's function is, because it is very hard to figure out what traits are adaptations for.

Are Selected Effects Functions Really Explanatory? Are they Normative?

There are two main arguments within this first group. The first tries to show that selected effects functions are not really explanatory. The second tries to show that they are not really normative. It would be fairly devastating if either of these were correct, since the best argument for the selected effects theory is that it accounts well for the explanatory and normative dimensions of function ascriptions. I will first examine the argument that functions are not explanatory and then the argument that they are not normative.

In 1975, Robert Cummins argued that selected effects functions are not explanatory, and that the only reason people think they are explanatory is because they are confused about what natural selection explains. In his view, selected effects theorists mistakenly believe that natural selection can explain why a given individual has a property (for example, why Amadi the zebra has stripes). But natural selection does not explain why a given individual has a property. Rather, it explains the present frequency of a property in a population.

Let me explain. Why does Amadi the zebra have stripes? It is not because natural selection gave him stripes. It is because of the interaction between his genes and his formative (including prenatal) environment. The reason Amadi has those genes, in turn, is because his parents had those genes and because of the mechanisms of heredity. The reason his parents had those genes is that their parents had those genes. When we go back far enough, we find that one of Amadi's very distant ancestors had a certain genetic mutation that caused that ancestor to have stripes, or proto-stripes. Natural selection is neither a proximal cause, nor a distal cause, for why Amadi has stripes. The selected effects theory seems to rest on an unfortunate confusion about how evolutionary biology works.

I believe that Cummins is getting at an important and interesting point. This is the distinction between what Sober (1984, 147–155) called "developmental" and "variational" explanations. Roughly, the first sort of explanation accounts for why an individual has a property by referring to the specific sequence of events that caused the individual to have that property (e.g., Amadi has stripes because his parents had stripes, and stripes are inherited, and his developmental context was conducive to the growth of stripes). The second sort of explanation accounts for the frequency of a given trait in a population. For example, why do zebras generally, or typically, have stripes (rather than being monocolored like horses)? Cummins and Sober believe that selection only explains this latter, population-level, fact, and not the former, individual-level fact. So, natural selection *can* explain why, say, most zebras have stripes. But it cannot explain why Amadi has stripes.

Some theorists reject Cummins' and Sober's conclusion (Neander 1988; see Sober 1995; Neander 1995a, b for further discussion). They argue that natural selection can explain not only why zebras generally have stripes, but also why Amadi the zebra has stripes. This has generated an ongoing discussion in the literature about what natural selection really explains (Walsh 1998; Nanay 2005; Helgeson 2015). This latter debate slides very quickly into subtle questions about the nature of causation and counterfactual reasoning in biology. If natural selection can explain why Amadi has stripes, then, plausibly, it does so by virtue of the correctness of the following counterfactual statement: if natural selection had acted differently, then Amadi the zebra would not have had stripes, but some other phenotype. It is not entirely clear how to evaluate this latter claim because it is not clear how we should identify Amadi in this counterfactual situation. In other words, how do we figure out which zebra is Amadi on those nearby possible worlds where there was no selection for stripes?

But I want to set those questions aside here. Let's suppose that Sober and Cummins are right that natural selection cannot explain why a given individual has some property. There is still an obvious sense in which functions are explanatory. It seems to me that when I say, "the function of zebra stripes is to deter flies," I am not trying to explain why this or that individual zebra (say, Amadi) has stripes. I am trying to explain why zebras, generally, have stripes, rather than something else. But Sober and Cummins agree that natural selection can explain that fact. So what is the point of the argument? In my view, Cummins was simply mistaken about what functions purport to explain.

Later, Cummins (2002, 164–6) developed a second line of attack. Consider a function statement like, "the function of the heart is to circulate blood," or "the function of the eye is to see." According to the selected effects theory, the statement "the function of the eye is to see," is only true if, at some point in the past, there was a population of organisms in which there was variation for having eyes. Some of these organisms, the ones with eyes, could see, and some, the ones without eyes, could not, and the former were selected over the latter. So far, so good. But Cummins says that typically, natural selection does not really work this way. It does not typically involve a contest between organisms that possess a given trait and organisms that entirely lack it. Rather, natural selection selects for slight improvements in functional ability. So again, the selected effects theorist is just confused about how natural selection works (also see Cummins and Roth 2010).

Let me give an example that I believe illustrates Cummins' point, if I understand him correctly. Consider the human eye in all of its complexity. Suppose we were to ask what the function of the *human* eye is. Presumably, we would want to know what it was selected for most recently. Let us suppose that the most recent and significant selection event took place roughly two million years ago with the emergence of the genus *Homo*. Specifically, unlike other great apes, the human eye exhibits a sharp color contrast between the sclera (the white part) and the iris (the colored part). Some theorists believe that this color contrast evolved by natural selection because it helped our early human ancestors follow each others' gaze (Tomasello et al. 2007). Let us suppose this is correct. Then it would seem to be correct to say that the function of the human eye (unlike, say, the ape eye) is to help us follow each others' gaze. The *human eye* was never selected for sight. That is, that is not what it was selected for over its closest living homologue (the primate eye). So, according to the selected effects theory, we should not say that the function of the human eye is to see, which is absurd.

Put this way, Cummins' argument seems to have a flaw. It would show, at best, that one function of the human eye is to help us follow each others' gaze. But that would not contradict the claim that *another* function of the human eye is to see. That is because, if one goes back far enough in time, one reaches a population of organisms, perhaps ancient flatworms, in which the selectionist account is probably right: some had eyes; some didn't have eyes; the ones with eyes were able to see; the ones without eyes were not able to see; the former were selected over the latter on that account; and our human eyes have descended from this ancient lineage. So if the selected effects theory is correct, then it would be correct to say *both* that a function of the human eye is to see (by virtue of the fact that it is an eye) *and* that a function of the human eye is to help us follow each others' gaze (by virtue of the fact that it exhibits a sharp color contrast between the sclera and iris). What is the problem?

Let me move on to a very different line of attack, which purports to show that the selected effects theory is not normative. This argument is due to Davies (2000, 2001, 2009). Remember, by "normativity," all I mean is that it is possible for a trait to have a function that it cannot perform. Something like malfunction is possible. Davies believes that this proposition involves a contradiction. Like Cummins, he thinks that the selected effects theorist is confused about how natural selection works.

Before I delve into his specific argument against normativity, let me explain the background of that argument. Davies is a proponent of the causal role theory of function, which I will discuss in Chap. 5. The causal role theory says, roughly, that the function of a trait consists in its contribution to some system level capacity that we are interested in. A function of the stomach is to break down food, because that capacity helps it to contribute to digestive system's ability to digest. The important point is that selected effects functions are a subset of causal role functions. If stripes have the selected effect function of warding off flies, then, at least at one point in the past, they must have had the causal role function of warding off flies. (Griffiths 1993 also makes this point.)

As a consequence of this (that selected effects functions are a subset of causal role functions), Davies thinks that selected effects functions are redundant or irrelevant. Why do we need them if the causal role theory lets us attribute functions to everything we want to attribute functions to, and more? I do not agree with Davies' conclusion here, since I think the causal role theory has problems of its own (see Sect. 5.2). Even if he is right that selected effects functions are a subset of causal role functions, that does not strike me as a good argument for rejecting the selected effects theory. It would be like saying that, since *justified true beliefs* are merely a subset of *true beliefs*, we should dispense with the justified true belief theory of knowledge and accept the true belief theory of knowledge instead.

Now, some people would argue that selected effects functions are not a subset of causal role functions, because they possess a special property that causal role functions lack. Namely, they can account for normativity in a way that the causal role theory cannot. Davies agrees that the causal role theory cannot account for normativity. His goal is to show that the selected effects theory cannot, either. So, he concludes, there is absolutely no advantage of accepting the selected effects theory, over accepting the causal role theory alone.

Here is, roughly, Davies' argument (see Davies 2001, 203; also Davies 2000). What is it for something to have the selected effects function of pumping blood? Well, it is to belong to the category of items that were selected for pumping blood. What sort of items are those? What properties, precisely, must an item possess in order to be a member of the category of items that were selected for pumping blood? Let's reflect on what was selected for. Hearts with four chambers were selected for over hearts with three chambers, over hearts with various sorts of hypertrophy, and over passive oxygen-diffusion systems. Those latter sorts of systems were, historically, selected *against*. So, in order for something to have the selected-effects function of pumping blood, it must have *just those properties* that were selected for (four chambers, absence of hypertrophy, etc.). Another way of putting it is that it must have just those properties that guaranteed successful pumping in ancestral environments.

Consider, now, someone born with a heart disease. Suppose the heart has only three chambers. This can result from a condition called tricuspid atresia. The selected effects theorist would like to say that such a heart *has* the function of pumping blood, but it *cannot* pump blood very well, so it is malfunctioning. But Davies says that a three-chambered heart is not actually a member of the category of things that were selected for pumping blood. (Technically, it is a member of the category of things that were selected against.) So a three-chambered heart doesn't have pumping as its function. Therefore, it is not malfunctioning. All of this is just a matter of understanding clearly how selection works.

I disagree with the way that Davies defines the selected effects theory of function. I think he defines the view too narrowly. If I understand him correctly, he thinks that a token trait has the (selected effects) function of pumping blood only if it is a member of the category of things that were selected for pumping blood. I define it differently. I would say that something has the selected effects function of pumping blood, *if it is related by descent* to a member of the category of things that were selected for pumping blood (namely, four-chambered hearts with no hypertrophy, and so on). I think this is a fairly standard construal of the selected effects theory. A three-chambered heart is related by descent to such a member, so it has the function of pumping blood. The reason that three-chambered hearts have the function of pumping blood is not that three-chambered hearts were ever selected for pumping blood, but because they stand in the right ancestor-descendent relations to something that was. Having a function, in my view, is a historical property like having royal blood. In order to have royal blood, you either have to be the king or queen, or you have to be related, by descent, to the king or queen.

Does the Selected Effects Theory Capture the Way Biologists Use the Term?

I will move on to the second major line of criticism against the selected effects theory. In this view, the selected effects theory is wrong because it contradicts the way that biologists actually use the term "function." The idea here is that biologists do not typically use the term "function" with any historical connotations. Instead, they seem to be more interested in current survival value (Horan 1989, 135; Amundson and Lauder 1994; Godfrey-Smith 1994, 351; Walsh 1996, 558; Schlosser 1998, 304; Wouters 2003, 658; Sarkar 2005, 18; Griffiths 2006, 3). Some of these authors refer to influential discussions by biologists to support their point. The ethologist Tinbergen (1963) has an important paper where he equates function with survival value and separates questions of function from questions of evolutionary history. The biologist Mayr (1961) made the same sort of point when he distinguished "functional" from "evolutionary" biology. Selected effects theorists seem determined to collapse the distinction between functional and evolutionary questions that biologists have worked very hard to set up.

We can get at the same point by looking closely at how biologists actually use the term "function" in day-to-day practice. Often, they seem to use the term without making any reference to history, but just by looking at present-day behavior. For example, sometimes, when biologists try to resolve competing hypotheses about a trait's function, they undertake careful observations of the trait's current-day behavior. They may even construct artificial environments or other experimental manipulations. But they do not generally look at a trait's history. That suggests that by "function," they are just referring to some feature of present-day behavior and not to the past.

Here is a case in point. There has been an ongoing debate about the function of eyespots on butterfly wings. Do the eyespots scare away predators? Or do they deflect predator attack away from the vital organs? When the entomologist Kathleen Prudic and her colleagues (Prudic et al. 2015) recently tried to figure out which hypothesis was correct, they collected careful field observations about how eyespots help butterflies. They observed that eyespots typically help butterflies by deterring attack away from the organs, and not by scaring off predators. As far as Prudic was concerned, that solved the question about function decisively. She and her colleagues did not seem to think that there was any additional question that had to be resolved in order to figure out its function, namely, something about its history.

There are at least two sorts of responses that selected effects theorists can give here. One response is to emphasize function pluralism. Most selected effects theorists are function pluralists. They think that biologists use "function" in different ways on different occasions. Sometimes biologists use "function" in a way that is closer to the fitness-contribution view, and sometimes they use it in a way that is closer to the selected effects view (see Sect. 5.3 for more discussion of function pluralism). In Sect. 3.1, I gave an example of a biologist, Tim Caro, who uses the term "function" in a more historical sense, and I also gave a large number of examples of biologists who, in explicitly discussing "function," have said that the function of something is whatever it was selected for. So, I believe that this criticism is not very relevant. Selected effects theorists can agree that sometimes biologists use the term "function" without talking about history, but other times they do.

I would actually go much further. This is the basis for a second sort of response. Even when biologists do not *explicitly* refer to selection when they attribute functions to traits, perhaps they *implicitly* do. The selected effects theory can be seen as an attempt to explicate (in the sense of rendering explicit) this implicit usage. In other words, biologists often use the notion of function with explanatory and normative connotations. The selected effects theory, in my view, provides the best account of these explanatory and normative connotations of function. So, when biologists use "function" with these explanatory and normative connotations, they are, I believe, implicitly appealing to selection, even if they do not realize that this is what they are doing. Consider the difference-making approach to causation. In this view, when we say that one event causes another we are implicitly making claims about nearby possible worlds, even if we do not do so explicitly, and even if we do not realize that this is what we are doing. I return to this line of criticism again in Sect. 4.3.

Are There Devastating Counterexamples to the Selected Effects Theory?

A third sort of criticism is that the selected effects theory is susceptible to various counterexamples. These are real or hypothetical cases that fall into one of two categories. In the first kind of case, the selected effects theory claims that an item does not have a function, but intuitively, it does. In the second sort of case, the selected effects theory claims that an item has a function, but intuitively, it does not.

The first kind of counterexample would imply that being selected for (with the appropriate amendments, such as being selected for in the recent past) is not necessary for having a function. Something can have the wrong sort of selection history but still possess a function. The second kind of counterexample would imply that being selected for is not sufficient for having a function. Something can have the right selection history but still lack a function. Of course, the effectiveness of this argument depends partly on what a theory of function is supposed to do (see Sect. 1.3), that is, how much a theory of function can depart from conceptual analysis of lay usage. But I will set that aside, and assume for the time being that a theory of function should, at the very least, provide a conceptual analysis of the way biologists use the term.

I will start with the first case, where something has the wrong history but still (intuitively) has a function. Consider a random mutation that benefits an organism on its first appearance. For example, consider the first ancestor of modern zebras that acquired a striped coat. Suppose that stripes, in this ancestor, deterred biting flies. Then, some philosophers, and I suppose some biologists, would say that, at that very moment, it came to have the *function* of deterring flies. So natural selection is not necessary for having functions (Ruse 1973; Walsh 1996).

I am not very impressed with this example. My intuition (for what it's worth) tells me that, in the absence of selection, the activity should count as a lucky benefit, and not as a function. Moreover, I think this intuition can be substantiated on principled grounds. Suppose I were abducted by a religious cult and that my captors decided to kill me. Suppose that as they prepared to kill me, they noticed a birthmark on my shoulder. Suppose my captors believed that God loves people with birthmarks, and so they let me go. I would not say that the birthmark has the function of deterring murderers. I would say it is a lucky benefit. The same is true, I claim, in the case of the first ancestor of modern zebras that had stripes. Why count it as a function, rather than a lucky accident?

Granted, there is a significant disanalogy here between zebra stripes and birthmarks. As one of the reviewers pointed out, there is a reliable counterfactual connection between zebra stripes and deterring flies. There is no reliable counterfactual connection between birthmarks and deterring cult members. I agree entirely with this point. My point is simply that most people who think about functions are willing to make a distinction between functions and lucky accidents. So if someone wishes to say that the very first appearance of stripes on a zebra had the function of deterring flies, then they should have to explain why they would classify it that way, rather than as a lucky accident.

The more interesting case is the second sort of counterexample, in which the selected effects theory would seem to confer a function on something that intuitively does not have one. The way to get these sorts of counterexamples going is to imagine a *non-biological* case where something like selection is happening. A classic example is Bedau's (1991) interesting case of evolution in clay crystals. He claims that clay crystals undergo a sort of natural selection, but that nobody in their right mind would say they have functions.

Clay crystals are formed by a "seed," which is a group of molecules that have a very specific pattern or structure. Over time, new layers of molecules are added to this seed, and they follow the same pattern. In other words, the pattern in the seed is something like a template for the pattern in the newer layers. Interestingly, populations of clay crystals exhibit something like variation, inheritance, and differential fitness, the classical ingredients for the "recipe" of natural selection (see Lewontin 1970; Godfrey-Smith 2009). I will consider each ingredient in turn.

First, variation. Different clay crystals have different characteristic structures. This is because there are differences in the patterns exhibited by each seed. Second, inheritance. As new layers of molecules are added to the initial seed, the crystal gets larger and larger until a piece finally breaks off. The new piece is a seed in its own right, which forms the basis for a new crystal. (We can count breaking as a kind of reproduction.) Interestingly, the new seed has the same pattern as its parent seed. We can count this similarity of parent and offspring as an instance of inheritance. Third, differential fitness. What is most interesting here is that the different sort of patterns that clay crystals exhibit can affect the rate at which they grow and the rate at which they "reproduce." That means that certain patterns can be copied more successfully than others. According to our recipe booklet, we have everything we need for natural selection to take place. So, suppose we have a certain pattern, P, that gets copied more reliably than some alternate pattern, P^*, such that the frequency of P increases over P^*. Should we say that P has a function? Intuitively, it seems like we should not. But why not?

Schaffner (1993, 383) developed a similar counterexample to illustrate the same point, which involves a machine he calls the "cloner." We can imagine different types of ball bearings, some of which are smooth and some are rough. Suppose they are all rolling down an incline with various gaps in it. The smooth ones roll faster and tend to successfully cross the gaps; the rough ones tend to fall into the gaps and perish. Suppose, moreover, that when a ball bearing successfully crosses the gap it enters a mechanical cloner device that spits out two identical bearings. Over time, the average smoothness of ball bearings would start to change. This is something like a selection process, but it seems counterintuitive to say that the smooth ball bearings have the function of jumping over gaps.

I don't think that Schaffner and Bedau have to insist that these are *bona fide* cases of natural selection to make their point. Even if we do not accept that this is really natural selection, they raise an important question. These examples seem to exhibit, at a certain level of abstraction, the characteristic features of natural selection, but none of them have functions for that reason. So, what's so special about natural selection that it should be a function-bestowing process?

I will focus on the clay crystals for now. There are three ways that the selected effects theorist could respond to this case. The first is to bite the bullet and simply admit that the clay crystals or ball bearings *do* have functions. This is not entirely unprincipled. For example, Sarkar (2005, 18) develops a theory of function in which the function of a trait consists in its contribution to the persistence of the entity in which it is contained ("broad-sense function"). He notes that his theory would allow physical (non-biological) structures to possess functions, but he welcomes this result on the grounds that there is no principled distinction between living and non-living matter (40). Although he is not discussing the clay crystals specifically, he would not find the implication problematic. Millikan (1993, 116) also bites the bullet and says that crystals have functions. Keep in mind that Millikan's view of function is meant to be fairly liberal in that she ascribes selected effects functions not only to biological traits but also to learned behaviors and artifacts (including linguistic items). As she puts it, "if crystals can have functions, as well as learned behaviors, artifacts, words, customs, etc., that is fine by me."

A second approach would be to emphasize that the selected effects theory is a theory of biological function, and as such, it only applies to biological entities (Garson 2011). For example, Neander's (1991) view, as it is developed there, seems to require a genetic system of inheritance and hence it seems to apply only to biological entities. This way of solving the problem might seem unsatisfying, but it is not clear why it should be unsatisfying. After all, most selected effects theorists are already comfortable with the idea that a good explication of function need only cover the biological case *rather than* the artifact realm. But if so, why not take this a step further and say that the theory of function is only meant to apply to the biological realm *rather* than the merely physical realm of clay crystals and ball bearings?

A third approach is to try to find something distinctive about the uncontroversial cases of function—zebra stripes, beating hearts—that the counterexamples lack. Bedau himself argues that the differentiating feature is the notion of *value*. Zebra stripes are good for zebras. But clay crystals do not have a good. In doing so, Bedau accepts an *evaluative* theory of function. He thinks when we attribute a function to an entity we are saying that the performance of the function is good for the entity (e.g., Bedau 1992, 801).

I hesitate to take this step with him, because it gets into complicated questions about the nature of value. Are values objective features of the natural world? Or are values subjective? Either way, Bedau's theory of function leads us into difficult waters. If values are objective components of the natural world, then we need to revise our basic metaphysical account of reality to reflect this. If values are subjective, and functions rely on value judgments, then it would seem that functions are somehow subjective, too. I have no idea where such a theory leads us but it would be interesting to explore. Other philosophers who embrace an evaluative theory of function include Sorabji (1964), Woodfield (1976, 122), Fulford (1999, 416) and McLaughlin (2001, 181).

Short of embracing the idea that functions are inherently value-laden, are there any other principles that we can appeal to that would exclude clay crystals from having functions? Perhaps another approach is to reflect on the idea, developed by Godfrey-Smith (1994) and Price (2001) and described in Sect. 3.2, that in order for a trait to have a function it is not enough that it contributes to its own differential reproduction. Rather, the trait must do so by contributing the fitness of a larger system (or a different system). Maybe we could somehow argue that the crystal merely contributes to its own differential persistence and not the persistence of a more inclusive or different system. Price (2001, 38) discusses the crystal case and makes this point. But it is not clear to me that this is a viable solution (as she seems to recognize). After all, each layer in a clay crystal *does* contribute to the persistence of the succeeding layers of the crystal, and not merely to its own persistence over time. So perhaps we can say that each layer of the crystal does have a function by virtue of the way it contributes to the persistence (or reproduction) of the crystal as a whole, or by virtue of the way it contributes to the persistence of the other layers.

Here is a proposal for how the selected effects theorist could avoid the implication that clay crystals (and ball bearings) have functions. We can insist that a trait comes to possess a function by virtue of the way it contributes to its differential reproduction within a *population* of like entities (Garson 2012, 460). Then we say that a bunch of ball bearings in a cloner machine is not a population. We also say that a bunch of replicating clay crystals is not a population either. Obviously, this solution will only be convincing if we have a good definition of what a population is.

So, what is a population? A problem here is that biologists (and philosophers of biology) have not discussed extensively the question of what a population is (for exceptions, see Millstein 2009; Godfrey-Smith 2009; Matthewson 2015). Does a set of entities constitute a population merely by virtue of spatial proximity? Must there be some sort of geographical isolation? Or does a set of entities become a population because of their characteristic patterns of interaction? For my purposes, I will define a population as a set of entities (of the same species or type) that affect each others' ability to survive or reproduce. A set of ants in the same colony is a population because of the way the ants affect each others' fitness, and not because they happen to be in the same locale. This is similar to the way that David Sloan Wilson defines the notion of a group for the purpose of distinguishing between individual and group selection (Wilson 1975; also see Sober and Wilson 1998, 92–98).

This maneuver, albeit vague, takes care of the cloner case. Nothing in Schaffner's account implies that the relative success of one ball bearing affects the relative success of another, either positively or negatively. It also seems to take care of the clay crystal case. If there is just a bunch of replicating crystals along the side of a creek, and some happen to replicate more quickly than others, they would not constitute a population if they do not affect each others' fitness.

That leaves me open to the following objection. Suppose we change the counterexample slightly and create one in which a bunch of clay crystals *do* affect each others' fitness? Suppose there are two types of crystals distinguished by their characteristic patterns *P* and *P**. Suppose *P*-type crystals and *P**-type crystals are growing along the side of a creek. Suppose the *P*-type crystals replicate much faster than the *P**-type crystals and start to crowd out the latter, or start to utilize a disproportionate number of environmental resources (namely, silicic acid molecules in the creek). Then the two types would form a single population of competing crystals. But in this case, I would be inclined to bite the bullet and simply say that the *P*-type crystal (or its components) *does* have a function, namely, whatever it does that manages to draw more silicic acid molecules to itself. My intuition, for whatever it is worth, tells me that the *P*-type crystals are life-like enough to attribute functions to their parts.

Does the Selected Effects Theory Impose Unacceptable Epistemic Burdens on Biologists?

A fourth and final sort of criticism is that the selected effects theory imposes severe epistemic burdens on biologists. The selected effects theory in its traditional form equates functions and adaptations. But it is really hard, and sometimes impossible, to figure out what something is an adaptation for, or whether it is an adaptation at all (Gould and Lewontin 1979; Lewontin 1998). So, if we accept the selected effects theory, it would be really hard for biologists to figure out what a trait's function is, or even if it has one (Amundson and Lauder 1994, 356; Schlosser 1998, 323–4; Weber 2005, 37; Wouters 2005, 144).

I suppose the implication here is that we make life easy for the biologists who wish to discover the functions of traits, and just define "function" in terms of some present-day benefit, or current causal roles. I sympathize with this motivation, but I do not think this is relevant to assessing the correctness of the theory as an explication of biological usage. If, according to our best analysis of the concept of function, functions are hard to discover, then so be it! As Socrates knew, our job as philosophers is not to make life easy for anyone.

Let me put the point slightly differently, and less flippantly. One of the jobs of the philosopher is to figure out what biologists mean by a certain term. That is, our role is to figure out the best analysis of what biologists are doing when the use a term, or what sort of deeper metaphysical commitments they express when they use a term. If it turns out that, on our best analysis, the biologists use the term "function" in a way that comes with severe epistemic burdens, then that is *their* problem, not ours. Perhaps philosophers have other, more prescriptive roles; for example, perhaps sometimes we should urge biologists, and other scientists, to stop using words in certain ways, and start using them in other ways. So, to be more cautious I should say that *to the extent* that we wish to explicate the implicit commitments of some biological usage this objection is not relevant.

3.4   A Generalized Selected Effects Theory

In this section, I will propose a version of the selected effects theory that I have
developed in several publications (Garson 2010, 2011, 2012, 2015), and which is
inspired by earlier treatments of function. In order to have a function, a trait must
have been selected for something. I take this as a starting point for my analysis,
because I think it neatly accounts for the explanatory and normative features of
function.

But what is selection? What sorts of selection processes are relevant for functions? One conventional way of interpreting this notion is to think of natural
selection operating over a population of reproducing organisms. Neander's (1991)
definition of function emphasizes this kind of selection. But, as I note above
(Sect. 3.2), some writers on function have said that the kind of selection relevant to
functions might operate over very different sorts of populations. Wimsatt (1972, 14)
interpreted selection broadly to refer to any process that involves "blind variation
and selective retention" (along the lines of Campbell 1960). Very importantly, this
sort of selection applies to behaviors acquired as a result of trial-and-error learning,
where different behaviors are tried out somewhat randomly and the successful ones
are repeated or reinforced. Millikan (1984, 28) also thought that trial-and-error
learning is a sort of selection process.

One of the first philosophers to note a connection between teleology and learning
history was Israel Scheffler (1959), though Mace (1935) also discusses this connection. As I noted in Chap. 2, Scheffler is mainly known for his nearly devastating
critiques of the goal-contribution approach to teleology. But Scheffler also developed a positive proposal in that paper. In Scheffler's view, what makes an entity or
process teleological is not its current-day behavior but its history, and specifically
its learning history.

He uses the following example. Suppose, in the past, an infant cried occasionally
whenever his mother left the room. Suppose his crying had the effect of securing
her return: she came back and comforted him. Now, as a result, the infant cries
whenever his mother leaves the room. For Scheffler, the infant's crying *now* (that is,
subsequent to the history of reinforcement) counts as teleological. What makes it
teleological is not its current-day benefits but the past sequence of events that
explain it:

> The apparent future-reference of a teleological description of this present interval is thus not
> to be confused with prediction [as in the goal directedness account] … Rather, the teleo-
> logical statement tells us something of the genesis of the present crying …. Such an account
> is perfectly compatible with normal causal explanation (269–270).

There are other writers on function who have expanded the relevant notion of
selection to include learning by trial-and-error as well as normal natural selection.
Millikan's attempt in this direction was described in Sect. 3.2. David Papineau
seems to want to apply the notion of selection to the differential reinforcement of
beliefs in one's cognitive system (Papineau 1987, 65–67; Papineau 1993, 44–48).

Godfrey-Smith (1992, 292) also suggested that the selected effects theory could apply to the differential reinforcement of learned behaviors. Griffiths (1993, 419) was concerned with expanding the notion of selection to include both biological and artifact functions. His idea was that an artifact can acquire a novel function by virtue of the fact that its designer *imagined* a number of possible designs and *selected* one of those virtual designs because of its (presumed) effects.

I am not endorsing the specific details of any of those approaches I have just cited. However, I endorse the general move to extend the selected effects theory outside of the context of natural selection operating over individual organisms. Of course, to say that some behaviors may acquire novel functions because of their learning history is not the same as embracing some sort of naïve behaviorism. There are many mechanisms that underlie the acquisition of novel behaviors and trial-and-error is only one amongst others (Kingsbury 2008).

Another example of a biological process that would fit this bill is antibody selection (Garson 2012; Garson Submitted for publication). Some scientists and philosophers have noted that there is a general sort of selection process that would seem to apply to the differential proliferation of antibodies in the bloodstream as well as to reproducing organisms. Each antibody has a specific shape, or conformation. When an antibody comes into contact with an antigen (foreign particle) with the same shape, the antibody is differentially replicated in the bloodstream. Hence, the mature configuration of antibodies in the adult bloodstream is a result of a selection-type process. For the background to this clonal theory of antibody production and its relation to function, see Garson (2012, 468). In response to cases like these, philosophers and scientists have attempted to explicate a very general, abstract, characterization of what a selection process is that would apply to natural selection, learning by trial-and-error, and antibody production (Darden and Cain 1989; Cziko 1995; Hull et al. 2001). These theorists were not explicitly concerned with the functions debate.

I wish to take this picture one step further. All of these cases—natural selection on organisms, the reinforcement of learned behaviors by trial-and-error, and the differential replication of antibodies—can be described within the framework that Millikan constructs because they all involve entities that *reproduce* (reproductively established families). But Millikan's view contains an important and somewhat arbitrary restriction. Millikan's approach does not apply to entities that do *not* reproduce, but that merely persist better or worse than others. However, we sometimes want to ascribe functions to such items. I think that some entities acquire functions by virtue of their differential persistence. I will first describe a case that I believe to be intuitively compelling, and then I will articulate a theory of function that is broad enough to include it. I will then differentiate my theory from Millikan's, and finally defend it from a set of counterexamples.

Consider the mature synaptic structure of the human brain. There are about 100 billion neurons in the brain, and in total there about 100 trillion synapses between them. Our uniqueness, as individual human beings, has a lot to do with the ways these neurons are connected to each other. There are at least three different sorts of processes that explain synaptic structure. Some of these synaptic connections are probably hard-wired, that is, specified by our genetic code. Other synaptic connections come about by virtue of the simple learning rule proposed by the psychologist Hebb (1949): "neurons that fire together, wire together." That is, when one neuron, *A*, causes an action potential in another neuron, *B*, then one or both neurons undergo some sort of structural change such *A* is more likely to activate *B* in the future. A third route is through synapse selection or synaptic pruning. In some cases, one neuron has connections with a large number of other neurons, and some of these connections are redundant. These synapses then undergo a sort of "competition," which results in some of the synapses being retained and others being eliminated. There is nothing speculative about synapse selection, and there are several well-documented cases of it. The most well-documented cases are from the rat's neuromuscular junction and from the formation of ocular dominance columns in the mammal's visual cortex. I have discussed these cases extensively in other places (Garson 2011, 2012, 2015, Chap. 7).

Synapse selection is only one sort of neural selection. There are two other sorts of neural selection we should consider. The second sort is *whole-neuron selection*. This takes place when entire neurons compete, as it were, for the resources needed to survive. This sort of whole neuron selection is an important part of the development of the spinal cord, and has been extensively documented (e.g., Deppmann et al. 2008). Finally, the neuroscientist Gerald Edelman (e.g., 1987) developed a much more speculative account of neural selection, which he called "neural group selection." He thought that selection takes place not only between synapses, and whole neurons, but also between entire *groups* of neurons. The view is something like group selection applied to the brain. I do not know of any direct evidence for this sort of neural group selection, but I also do not know of any evidence against it. At any rate, even if one is skeptical about the reality of this sort of neural group selection, there is no basis for skepticism about the power of synapse selection and whole neuron selection.

I think that neural selection, on whatever level it occurs, is relevant to functions. I am going to focus on the case of synapse selection in particular, because it is easy to understand and it is uncontroversial that it takes place. I think that synapse selection can give rise to new functions in the brain. But there is nothing like differential replication taking place. There is no sense in which a given synapse reproduces itself. Rather, synapses merely persist better or worse than others. I think that if a synapse outcompetes another synapse then it acquires a new function. Its function is to do whatever it did that led to its differential persistence. This is a significant step beyond even Millikan's fairly liberal account of selection.

Here is a definition of function that I endorse. I call it the "generalized selected effects theory of function" (Garson 2010, 2011, 2012, 2015). It says that the function of a trait consists in the activity that led to its differential *reinforcement* or its differential *reproduction* in a biological population. The first part of the disjunction, the "differential reinforcement" part, is meant to include various forms of neural selection where there is no replication. The second part, "differential reproduction," is meant to cover the more traditional sort of natural selection (over organisms) as well as some of the more extended cases such as antibody selection. The third part, "in a biological population," is meant to exclude some of the counterexamples I raised in the last section, such as Bedau's clay crystals or Schaffner's ball bearings.

The view that is probably closest in spirit to my own is Bouchard (2013), who emphasizes the importance of differential persistence in thinking about functions. However, his view is a forward-looking one while mine is historical; moreover, he was more concerned about making sense of functions in ecology and I am, here, more concerned with neuroscience. Other theorists, such as Sarkar (2005, 18), have emphasized the idea that something comes to have a function by virtue of the way it contributes to the persistence of some larger entity. My view differs in that it insists that there must be something like selection taking place. Something can come to possess a function because it contributes to an entity's ability to persist better, or longer, than some other entity within the same population.

The main argument for this generalized selected effects theory is that it satisfies the same desiderata as the conventional selected effects view, but without seemingly arbitrary restrictions. It does not restrict functions to entities that reproduce. Like the traditional selected effects view, it insists that selection is essential to functions. After all, it is the reference to selection that excludes the Boorse-type counterexamples. It also provides a very natural explanation for how the brain generates new functions on a continuing basis. In Sect. 5.3 I will also describe how it leads to a new form of function pluralism.

There are various concerns one might have with the liberal theory. First, this generalized selected effects theory, like the traditional theory, defines function in terms of selection history. So, anybody who does not accept the traditional selected effects theory will probably not accept this generalized form either. I believe that I have adequately defended the traditional theory from common criticisms (in the previous section).

Another concern is that the generalized selected effects theory rests on a highly misleading analogy between natural selection and neural selection. (I thank Gerhard Schlosser for bringing this point to my attention.) They are very different from one another, for a number of reasons. One reason is that natural selection involves reproduction and inheritance, and this feature gives natural selection such an extraordinary power to generate novel, adaptive phenotypes. Neural selection does not possess this feature. In other words, the generalized theory only seems plausible because it rests on a misleading analogy between natural selection and other, superficially similar, processes.

However, my argument for the generalized selected effects theory is not that neural selection is similar to natural selection. I agree that that would be a pretty weak argument, as arguments by analogy tend to be. My argument is that the generalized selected effects theory satisfies the exact same desiderata as the traditional selected effects theory, and it does away with arbitrary restrictions. At the very least, then, anyone who accepts the traditional selected effects theory should accept the generalized selected effects theory.

Another concern here is that the generalized selected effects theory is *too* liberal. Won't it open the floodgates to a host of new counterexamples? That is a troubling prospect, particularly because we rejected Wright's theory on precisely that account. But I do not think that it creates many new counterexamples, over and above the sort that afflict the traditional selected effects theory (and which I described in the last section). The reference to *selection* excludes the Boorse-type counterexamples, and the reference to *populations* excludes the cloner-type counterexamples. Which new counterexamples might we worry about?

One might worry that there can be differential persistence without functions. For example, suppose there are a bunch of rocks on a beach that differ in their hardness (this counterexample is adapted from Kingsbury 2008). The soft rocks will wear down more quickly than the hard rocks. So there is something like differential persistence, but no new functions. This is not a counterexample to my view, because a bunch of rocks on a beach does not constitute a population. The reason it does not constitute a population is because, as noted in the previous section, the rocks do not affect each others' fitness (survival prospects). So far, so good.

But imagine that we change the counterexample somewhat. (I thank Karen Neander for this interesting challenge.) Imagine that all the rocks are piled on top of one another so that, when the waves come in, they rub against each other. As a consequence, they start to wear each other down. Now, it looks like we have something like a population. There are a bunch of individuals that affect the others' chances of persistence. Now, if some of those rocks are harder than others, and hence persist for longer than others, then we have a case of differential persistence within a population. But it seems counterintuitive to attribute functions to the hardness of rocks. Moreover, this is a counterexample that does not arise for the traditional selected effects theory. That is because rocks do not reproduce or replicate. So my theory seems to admit counterexamples that the traditional theory does not.

I admit I am not entirely sure how to resolve this sort of counterexample. I would say, however, that the exact same three options are open to me that are open to the traditional selected effects theorist (and which I enumerated in the last section). First, I could bite the bullet and say that hardness in rocks *can* be a function. That does not sound like a very happy resolution. Second, I could argue that the case is not a legitimate counterexample because I am only developing a theory of biological functions, and rocks are not biological entities. Maybe the rocks have some sort of function, but it is not a biological function. That sounds a little bit more promising, but also a little bit ad hoc.

Third, I could try to argue that there is some other differentiating feature that makes a population of rocks very unlike a population of, say, synapses, and that is relevant to functions. For example, I could try to argue that an individual rock cannot have a function because it is an uncontained container. In other words, though the individual rock promotes its own differential persistence (over other rocks), it does not do so by contributing to the fitness of a larger (or different) system. Hence, a rock that differentially persists over others is more like a cancer cell multiplying more rapidly than normal cells. The cell, as a whole, does not have a function because it is not contributing to any other system, but only to its own proliferation.

Another sort of approach along these lines would be to suggest that the reasons the properties of rocks do not have functions is that rocks do not have the right sort of inner complexity. As Bence Nanay suggested to me, rocks are not self-maintaining systems, while organisms are, and maybe that has to do with why we attribute functions to parts of organisms and not to rocks. Such a theory would bring my view of function extremely close to another theory, the organizational (or systems-theoretic) approach to function, which I discuss in Sect. 6.2.

A different sort of solution would be to suggest that a pile of rocks is not really a population, contrary to initial appearance. Matthewson (2015) recently argued that bona fide populations exhibit high degrees of what he calls "linkage." In other words, in a biological population, changes in the fitness of one individual bring about changes in the fitness of many others, and not just a handful of its immediate neighbors. Perhaps a pile of rocks does not exhibit a high enough degree of linkage to constitute a real population. At any rate, my point is not to offer some definitive resolution of this puzzle, but merely to outline various possible forms that such a solution might take.

References

Amundson, R., & Lauder, G. V. (1994). Function without purpose: The uses of causal role function in evolutionary biology. *Biology and Philosophy, 9*, 443–469.

Ayala, F. J. (1968). Biology as an autonomous science. *American Scientist, 56*, 207–221.

Ayala, F. J. (1970). Teleological explanations in evolutionary biology. *Philosophy of Science, 37*, 1–15.

Bedau, M. (1991). Can biological teleology be naturalized? *Journal of Philosophy, 88*, 647–655.

Bedau, M. (1992). Where's the good in teleology? *Philosophy and Phenomenological Research, 52*, 781–802.

Boorse, C. (1976). Wright on functions. *Philosophical Review, 85*, 70–86.

Bouchard, F. (2013). How ecosystem evolution strengthens the case for function pluralism. In P. Huneman (Ed.), *Function: Selection and mechanisms* (pp. 83–95). Dordrecht: Springer.

Brandon, R. N. (1981). Biological teleology: Questions and explanations. *Studies in History and Philosophy of Science A, 12*, 91–105.

Buller, D. J. (1998). Etiological theories of function: A geographical survey. *Biology and Philosophy, 13*, 505–527.

Campbell, D. T. (1960). Blind variation and selective survival as a general strategy in knowledge processes. In M. C. Yovits & S. Cameron (Eds.), *Self-organizing systems* (pp. 205–231). New York: Pergamon Press.

Caro, T., et al. (2014). The function of zebra stripes. *Nature Communications, 5*, 3535.

Crow, J. F. (1979). Genes that violate Mendel's rules. *Scientific American, 240*(2), 134–146.

Cummins, R. (1975). Functional analysis. *Journal of Philosophy, 72*, 741–765.

Cummins, R. (2002). Neo-teleology. In A. Ariew, R. Cummins, & M. Perlman (Eds.), *Functions: New essays in the philosophy of psychology and biology* (pp. 157–172). Oxford: Oxford University Press.

Cummins, R., & Roth, M. (2010). Traits have not evolved to function the way they do because of a past advantage. In F. J. Ayala & R. Arp (Eds.), *Contemporary debates in philosophy of biology* (pp. 72–85). Malden, MA: Blackwell.

Cziko, G. (1995). *Without miracles: Universal selection theory and the second darwinian revolution*. Cambridge: MIT Press.

Darden, L., & Cain, J. A. (1989). Selection type theories. *Philosophy of Science, 56*, 106–129.

Davies, P. S. (2000). Malfunctions. *Biology and Philosophy, 15*, 19–38.

Davies, P. S. (2001). *Norms of nature: Naturalism and the nature of functions*. Cambridge, MA: MIT Press.

Davies, P. S. (2009). Conceptual conservatism: The case of normative functions. In U. Krohs & P. Kroes (Eds.), *Functions in biological and artificial worlds* (pp. 127–146). Cambridge, MA: MIT Press.

Deppmann, D., et al. (2008). A model for neuronal competition during development. *Science, 320*, 369–373.

Doolittle, W. F. (1989). Hierarchical approaches to genome evolution. *Canadian Journal of Philosophy, 14*(suppl), 101–133.

Doolittle, W. F. (2013). Is junk DNA bunk? A critique of ENCODE. In: *Proceedings of the National Academy of Sciences*. doi: 10.1073/pnas.1221376110

Edelman, G. M. (1987). *Neural Darwinism: The theory of neuronal group selection*. New York: Basic Books.

Elliott, T. A., Linquist, S., & Gregory, T. R. (2014). Conceptual and empirical challenges to ascribing functions to transposable elements. *American Naturalist, 184*, 14–24.

Fulford, K. W. M. (1999). Nine variations and a coda on the theme of an evolutionary definition of dysfunction. *Journal of Abnormal Psychology, 108*, 412–420.

Garson, J. (2010). Schizophrenia and the dysfunctional brain. *Journal of Cognitive Science, 11*, 215–246.

Garson, J. (2011). Selected effects functions and causal role functions in the brain: The case for an etiological approach to neuroscience. *Biology and Philosophy, 26*, 547–565.

Garson, J. (2012). Function, selection, and construction in the brain. *Synthese, 189*, 451–481.

Garson, J. (2015). *The biological mind: A philosophical introduction*. London: Routledge.

Garson, J. (Submitted for publication). How to be a function pluralist. *British Journal for the Philosophy of Science*.

Godfrey-Smith, P. (1992). Indication and adaptation. *Synthese, 92*, 283–312.

Godfrey-Smith, P. (1994). A modern history theory of functions. *Nous, 28*, 344–362.

Godfrey-Smith, P. (2009). *Darwinian populations and natural selection*. Oxford: Oxford University Press.

Gould, S. J., & Lewontin, R. (1979). The Spandrels of San Marco and the panglossian paradigm. *Proceedings of the Royal Society of London, 205*, 281–288.

Gould, S. J., & Vrba, E. S. (1982). Exaptation: A missing term in the science of form. *Paleobiology, 8*, 4–15.

Griffiths, P. E. (1992). Adaptive explanation and the concept of a vestige. In P. Griffiths (Ed.), *Trees of life: Essays in philosophy of biology* (pp. 111–131). Dordrecht: Kluwer.

Griffiths, P. E. (1993). Functional analysis and proper function. *British Journal for the Philosophy of Science, 44*, 409–422.

Griffiths, P. E. (2006). Function, homology, and character individuation. *Philosophy of Science, 73*, 1–25.

Hebb, D. O. (1949). *The organization of Behavior*. New York: Wiley.

Helgeson, C. (2015). There is no asymmetry of identity assumptions in the debate over selection and individuals. *Philosophy of Science, 82*, 21–31.

Horan, B. (1989). Functional explanations in sociobiology. *Biology and Philosophy, 4*, 131–158.

Hull, D. L., Langman, R. E., & Glenn, S. S. (2001). A general account of selection: Biology, immunology and behavior. *Behavioral and Brain Sciences, 24*, 511–527.

Kingsbury, J. (2008). Learning and selection. *Biology and Philosophy, 23*, 493–507.

Kraemer, D. M. (2014). Revisiting recent etiological theories of functions. *Biology and Philosophy, 29*, 747–759.

Krohs, U., & Kroes, P. (Eds.). (2009). *Functions in biological and artificial worlds*. Cambridge, MA: MIT Press.

Larison, B., et al. (2015). How the zebra got its stripes: A problem with too many solutions. *Royal Society Open Science*. DOI: 10.1098/rsos.140452.

Lennox, J. G. (1993). Darwin was a teleologist. *Biology and Philosophy, 8*, 409–421.

Lennox, J. G., & Kampourakis, K. (2013). Biological teleology: The need for history. In K. Kampourakis (Ed.), *The philosophy of biology: A companion for educators* (pp. 421–454). Dordrecht: Springer.

Lewens, T. (2004). *Organisms and artifacts: Design in nature and elsewhere*. Cambridge, MA: MIT Press.

Lewontin, R. C. (1970). The units of selection. *Annual Review of Ecology and Systematics, 1*, 1–18.

Lewontin, R. C. (1998). The evolution of cognition: Questions we will never answer. In D. Scarborough & S. Sternberg (Eds.), *An invitation to cognitive science, vol 4: Methods, models, and conceptual issues* (2nd ed., pp. 107–132). Cambridge, MA: MIT Press.

Lorenz, K. (1966) [1963]. *On Aggression*. New York: Harcourt, Brace & World.

Mace, T. A. (1935). Mechanical and teleological causation. *Proceedings of the Aristotelian Society, S14*, 22–45.

Maley, C. J., & Piccinini, P. (Submitted for publication). A unified mechanistic account of teleological functions for psychology and neuroscience. In D. Kaplan (Ed.), *Integrating psychology and neuroscience: Prospects and problems*. Oxford: Oxford University Press.

Matthewson, J. (2015). Defining paradigm darwinian populations. *Philosophy of Science, 82*, 178–197.

Maynard Smith, J. (1990). Explanation in biology. In D. Knowles (Ed.), *Explanation and its Limits* (pp. 65–72). Cambridge: Cambridge University Press.

Mayr, E. (1961). Cause and effect in biology. *Science, 134*, 1501–1506.

McLaughlin, P. (2001). *What functions explain: Functional explanation and self-reproducing systems*. Cambridge: Cambridge University Press.

Millikan, R. G. (1984). *Language, thought, and other biological categories*. Cambridge, MA: MIT Press.

Millikan, R. G. (1989a). In defense of proper functions. *Philosophy of Science, 56*, 288–302.

Millikan, R. G. (1989b). An ambiguity in the notion 'function'. *Biology and Philosophy, 4*, 172–176.

Millikan, R. G. (1993). *White queen psychology and other essays for Alice*. Cambridge, MA: MIT Press.

Millikan, R. G. (1999). Wings, spoons, pills, and quills: A pluralist theory of function. *Journal of Philosophy, 96*, 191–206.

Millikan, R. G. (2002). Biofunctions: Two paradigms. In A. Ariew, R. Cummins, & M. Perlman (Eds.), *Functions: New essays in the philosophy of psychology and biology* (pp. 113–143). Oxford: Oxford University Press.

Millstein, R. L. (2009). Populations as individuals. *Biological Theory, 4*, 267–273.

Nagel, E. (1977). Teleology revisited: Goal directed processes in biology and functional explanation in biology. *Journal of Philosophy, 74*, 261–301.

Nanay, B. (2005). Can cumulative selection explain adaptation? *Philosophy of Science, 72*, 1099–1112.

Neander, K. (1983). *Abnormal psychobiology*. Dissertation, La Trobe.

Neander, K. (1988). What does natural selection explain? Correction to Sober. *Philosophy of Science, 55*, 422–426.

Neander, K. (1991). Functions as selected effects: The conceptual analyst's defense. *Philosophy of Science, 58*, 168–184.

Neander, K. (1995a). Pruning the tree of life. *British Journal for the Philosophy of Science, 46*, 59–80.

Neander, K. (1995b). Explaining complex adaptations: A reply to Sober's 'Reply to Neander'. *British Journal for the Philosophy of Science, 46*, 583–587.

Papineau, D. (1987). *Reality and representation*. Oxford: Blackwell.

Papineau, D. (1993). *Philosophical naturalism*. Oxford: Blackwell.

Piccinini, G. (2015). *Physical computation: A mechanistic account*. Oxford: Oxford University Press.

Preston, B. (1998). Why is a wing like a spoon? A pluralist theory of function. *Journal of Philosophy, 95*, 215–254.

Preston, B. (2003). Of marigold beer: A reply to Vermaas and Houkes. *British Journal for the Philosophy of Science, 54*, 601–612.

Price, C. (2001). *Functions in mind: A theory of intentional content*. Oxford: Oxford University Press.

Prudic, K. L., et al. (2015). Eyespots deflect predator attack increasing fitness and promoting the evolution of phenotypic plasticity. *Proceedings of the Royal Society B, 282*, 201415.

Ruse, M. E. (1973). A reply to Wright's analysis of functional statements. *Philosophy of Science, 40*, 277–280.

Sarkar, S. (2005). *Molecular models of life*. Cambridge, MA: MIT Press.

Schaffner, K. (1993). *Discovery and explanation in biology and medicine*. Chicago: University of Chicago Press.

Scheffler, I. (1959). Thoughts on teleology. *British Journal for the Philosophy of Science, 9*, 265–284.

Schlosser, G. (1998). Self-re-production and functionality: A systems-theoretical approach to teleological explanation. *Synthese, 116*, 303–354.

Schwartz, P. H. (1999). Proper function and recent selection. *Philosophy of Science, 66*, S210–S222.

Schwartz, P. H. (2002). The continuing usefulness account of proper function. In A. Ariew, R. Cummins, & M. Perlman (Eds.), *Functions: New essays in the philosophy of psychology and biology* (pp. 244–260). Oxford: Oxford University Press.

Sherrington, C. S. (1906). *The integrative action of the nervous system*. New Haven: Yale University Press.

Sober, E. (1984). *The nature of selection*. Chicago: University of Chicago Press.

Sober, E. (1995). Natural selection and distributive explanation: A reply to Neander. *British Journal for the Philosophy of Science, 46*, 384–397.

Sober, E., & Wilson, D. S. (1998). *Unto others: The evolution and psychology of unselfish behavior*. Cambridge, MA: Harvard University Press.

Sorabji, R. (1964). Function. *Philosophical Quarterly, 14*, 289–302.

Tinbergen, N. (1963). On aims and methods of ethology. *Zeitschrift für Tierpsychologie, 20*, 410–433.

Tomasello, M., et al. (2007). Reliance on head versus eyes in the gaze following of great apes and human infants: The cooperative eye hypothesis. *Journal of Human Evolution, 52*, 314–320.

Vermaas, P. E., & Houkes, W. (2003). Ascribing functions to technical artefacts: A challenge to etiological accounts of functions. *British Journal for the Philosophy of Science, 54*, 261–289.

Walsh, D. M. (1996). Fitness and function. *British Journal for the Philosophy of Science, 47*, 553–574.

Walsh, D. M. (1998). The scope of selection: Sober and Neander on what natural selection explains. *Australasian Journal of Philosophy, 76*, 250–264.

Weber, M. (2005). *Philosophy of experimental biology*. Cambridge: Cambridge University Press.

Williams, G. C. (1966). *Adaptation and natural selection: A critique of some current evolutionary thought*. Princeton: Princeton University Press.

Wilson, D. S. (1975). A theory of group selection. *Proceedings of the National Academy of Sciences USA, 72*, 143–146.

Wimsatt, W. C. (1972). Teleology and the logical structure of function statements. *Studies in the History and Philosophy of Science, 3*, 1–80.

Wimsatt, W. C. (2002). Functional organization, analogy, and inference. In A. Ariew, R. Cummins, & M. Perlman (Eds.), *Functions: New essays in the philosophy of psychology and biology* (pp. 173–221). Oxford: Oxford University Press.

Woodfield, A. (1976). *Teleology*. Cambridge: Cambridge University Press.

Wouters, A. (2003). Four notions of biological function. *Studies in the History and Philosophy of Biological and Biomedical Sciences, 34,* 633–668.

Wouters, A. G. (2005). The functional perspective in organismic biology. In T. A. C. Reydon & L. Hemerik (Eds.), *Current themes in theoretical biology* (pp. 33–69). Dordrecht: Springer.

Wouters, A. G. (2013). Biology's functional perspective: Roles, advantage, and organization. In K. Kampourakis (Ed.), *The philosophy of biology: A companion for educators* (pp. 455–486). Dordrecht: Springer.

Chapter 4
Function and Fitness

Abstract This chapter focuses on the fitness-contribution theory of function, which holds, roughly, that the function of a trait consists in its typical contribution to the fitness of the organisms that possess it. I begin by surveying several different theories within this family, and I show why any plausible version must include a statistical element. I then pose three questions that any proponent of the fitness-contribution theory must answer. First, is fitness a relative notion? When one says a trait "contributes to fitness," is one saying it contributes to fitness better than some alternative? If so, when we attribute a function to a trait, how do we specify the relevant alternatives? Second, is fitness relative to specific environments? If so, then when we attribute a function to a trait, how do we specify the relevant environments? Third, what precisely must a trait contribute to in order to have a function? Is it survival, reproduction, inclusive fitness, or something else? I then critically assess a major argument in its favor, namely, that it coheres well with the way biologists actually use the term. I consider three different interpretations of this claim and I argue that it does not, in fact, provide an advantage over the selected effects theory in this regard. I close by considering how well it satisfies the adequacy conditions set out in Chap. 1. Theorists disagree about whether the fitness-contribution theory can make sense of the explanatory and normative aspects of function and I survey those disagreements.

Keywords Fitness-contribution functions · Boorse functions · Propensity functions · Biostatistical functions · Forward-looking functions

4.1 Functions and Statistical Normalcy

In 1964, the philosopher of science John Canfield proposed a simple analysis of function that stood in stark contrast to the goal directedness accounts of teleology that were then popular. He attempted to capture both artifact functions and biological functions in terms of their current-day benefit. My air purifier benefits me by helping me breathe clean air. The zebra's stripes benefit it by deterring

© The Author(s) 2016
J. Garson, *A Critical Overview of Biological Functions*,
Philosophy of Science, DOI 10.1007/978-3-319-32020-5_4

disease-carrying flies. Simply put, in his analysis, the function of something is just the benefit it brings.

Can we say anything more about this benefit, when it comes to the biological realm? After all, to benefit someone is typically to do something good for that person. So the language of benefits is most at home in the realm of human values and needs. We might wish to resist importing the language of benefits and values into the way we talk about the natural world.

Canfield thought that, in the context of biology (e.g., the function of the zebra's stripes), "benefit" should be defined in terms of fitness. Specifically, to say that a function of a trait, T, is to do F means that the probability that creatures with T survive or reproduce is greater than the probability that creatures without T survive or reproduce. So, even though Canfield articulated his view in the language of benefits, he is the founder of the tradition that defines "function" in terms of current-day fitness contributions (e.g., Lehman 1965; Ruse 1971; Boorse 1976; Bechtel 1986; Bigelow and Pargetter 1987; Horan 1989; Wouters 1995, 2005, 2013; Walsh 1996; Walsh and Ariew 1996; Lewens 2004; Kraemer 2013; Garson and Piccinini 2014). He was also very explicit in his view that "function" should be defined by using the concepts of evolutionary biology (Canfield 1966). This represented a significant shift in focus away from the preoccupation with cybernetics and homeostatic devices.

This chapter will describe the development of this fitness-contribution theory of functions. Rather than offering a straight historical narrative, however, my exposition will proceed dialectically, that is, by presenting the theory with a series of problems and showing how the theory could, or even must, be articulated so as to avoid those problems.

The first section (Sect. 4.1) shows why the theory is best articulated as a *statistical* theory of function, and will survey several variations on that basic theme. The second section (Sect. 4.2) asks three questions any theory of this form should answer. First, if the notion of a "contribution to fitness" is *relative*, that is, if, when we say that a trait contributes to fitness, we mean that it contributes to fitness *better than* some real or imagined alternative, how should we specify the relevant alternatives? Second, if fitness is relative to an environment (such that a trait can be more fit in one environment and less fit in another) then how do we specify the relevant environment? Third, what *exactly* must a trait contribute to in order to have a function? Is it survival, reproduction, fitness, or something else? Most writers have been vague on that score. The final section (Sect. 4.3) steps back to provide a more global assessment of the theory. Why should we accept the theory, and can it satisfy our three desiderata: the function/accident distinction, the explanatory aspect of functions, and the normative aspect of functions?

The fitness-contribution theory is best framed as a statistical theory about trait *types*, rather than a theory about tokens. A trait token can benefit an individual on a single occasion, or even on a number of occasions, but that would not give it a function. A soldier's pocket Bible might stop a bullet, but that is not its function. My nose helps me by supporting my glasses, but that is not its function either. So, at least one natural interpretation of Canfield's formula cannot be right, namely, one that holds that the function of a trait token is just whatever that token does that contributes to the fitness of the organism that possesses it. Such a formulation would violate the distinction between functions and accidental benefits, which has been a staple of the functions literature since its inception (Frankfurt and Poole 1965; Wright 1972). Of course, we could say that the pocket Bible "functions as" a bullet-stopper on that occasion, or that my nose "functions as" a glasses-holder, but I believe this is a different concept than the biological notion of function (Wright 1973, 143). I will return to this "functions as" notion in Sect. 5.2.

The best way to avoid this problem is to define the function of a trait in terms of its *statistically significant* contribution to fitness, that is, the contribution that tokens of that type typically make, in a species or population, to some measure of fitness. The reason my heart has the function of circulating blood and not making beating sounds is because blood circulation is what the heart typically does in the species that benefits the individuals that have it. Just because, occasionally, the heart's beating sounds benefit some individual (say, by alerting a doctor to a life-threatening illness), that does not give it the function of doing so. (Of course, if that situation becomes very typical in the future, say, the heart's beating sounds frequently alert doctors to potential heart problems, then, at that time, one function of the heart will be to make heart sounds. Perhaps in some parts of the world it already has.)

The idea that functions have to do with statistically-typical contributions to fitness is embodied in Christopher Boorse's influential theory of function (e.g., Boorse 1976, 1977, 2002). Boorse defines the *physiological* function of a trait in terms of what is, in some sense, typical for the species or population. As I noted in Sect. 2.1, there are some interpretive problems in classifying Boorse's view. At the most general level, he holds that the function of a trait is just its contribution to a goal of a broader system (1976, 78). Boorse explicitly defined the goal of a system in cybernetic terms, that is, in the style preferred by theorists such as Sommerhoff and Nagel (described in Sect. 2.2), where the goal of a system is that toward which the system is "directively organized." Theoretically, then, Boorse's view is a version of the goal-contribution theory of function. It is neutral between organisms and artifacts. An organism can have many different goals, but he thought that, when we are talking about the context of physiology and of biomedicine, the relevant goals were survival and reproduction. So, in that context, a function of a trait is, roughly, its species-typical contribution to fitness (1976, 84; 1977, 555). In other words, Boorse's goal-contribution theory, and the fitness-contribution theory, yield the same conclusions when we restrict the scope of the theory to physiology and medicine.

So, what *exactly* does the notion of a "typical contribution to fitness" mean in this context? Millikan (1984, 34) raised a potential counterexample to the view when she noted that some functions are performed very *atypically*. For example, the function of sperm is to fertilize ova, but the probability that any given sperm will actually fulfill that function is vanishingly small. In what sense is it typical for a sperm to fertilize ova?

This is only a serious counterexample for Boorse's view if we interpret the notion of "typicality" in a rather strained way, namely, as the unconditional probability that a given token of a trait performs the activity in question. In this sense, it is highly improbable that any given sperm will fertilize an ovum. Although Boorse was not entirely clear on how he wished this notion of typicality to be understood, he suggested a different, more natural interpretation of typicality, where typicality is understood as a conditional probability. Specifically, we should say that a trait *T* has a function *F* when the conditional probability that *T* does *F* at any given moment, given that it does *something* to contribute to survival and reproduction at that moment, is very high (Boorse 2002, 92–3). The probability that any token sperm is fertilizing an ovum, given that it is doing *something* that contributes to survival and reproduction, is very high.

Neander, however, raised a problem for Boorse's way of interpreting the typicality condition. (Neander did not raise it in print, but Boorse (2002, 93, *fn.* 34) acknowledged that she raised the difficulty in personal communication). We can call it the "problem of multi-functional traits." Suppose a trait, *T*, has two different functions, *F1* and *F2*. Suppose *F1* is performed much more commonly than *F2*. So, the probability that *T* does *F1*, given that it is contributing to fitness, is high, but the probability that *T* does *F2*, given that it is contributing to fitness, is low. Then *F2* would not be a function of *T*. But a simple example shows that this is counterintuitive. Consider the brain's medulla, which performs a number of vital functions. One function (*F1*) is to contribute to breathing, and it performs this very commonly. Another function (*F2*) is to contribute to the gag reflex, and it performs this very rarely (relative to *F1*). So Boorse's recommended modification implies that the medulla does not have the function of initiating the gag reflex, and that sounds wrong.

Garson and Piccinini (2014) amend Boorse's view of function to resolve this problem. They argue that a trait *T* has a function *F* when the conditional probability that *T* does *F*, given that it is doing something to contribute to fitness, is *non-negligible*, rather than very high. They believe that this formulation avoids Neander's problem of multi-functional traits, since the probability that the medulla contributes to gagging, given that it is doing something helpful, is non-negligible. It also preserves the function/accident distinction. Presumably, the probability that the heart contributes to survival by making sounds, given that it is doing *something* to contribute to fitness, *is* negligible, since, in the overwhelming majority of cases in which the heart contributes to fitness, it does so by pumping blood. Garson and Piccinini note, however, that the distinction between negligible and non-negligible contributions to fitness is vague and partly contextually-determined.

There are other formulations of the fitness-contribution theory. I think most of them, at least implicitly, incorporate a statistical element. One variation of Boorse's view is the idea that the function of a trait should be defined in terms of what I will call its "selection-proneness." That is, *T* has *F* if *T*'s doing *F* makes *T* prone to be selected over some alternative, *T**. This is an explicitly relational view, where the function of a trait is defined in terms of its advantage over some real or hypothetical alternatives. (I will develop a problem for this view in the next section.) I believe that Ruse (1971, 1973a, b), Bigelow and Pargetter (1987), Walsh (1996) and Walsh and Ariew (1996) all endorse some version of this idea, as I will presently describe. One way of thinking about the difference between Boorse's view and these other views is that on Boorse's view, logically speaking, a trait could have a function in a world where natural selection does not take place (for example, a world where all of the species pop into existence at once in their present form). On these latter views, where function has to do with selection-proneness, traits could not have functions in a world without natural selection.

Ruse (1971, 92) proposed an idea similar to Canfield's. His view was that, in order for T to have the function F, F must be an adaptation. He went on to say that an effect of a trait is an adaptation when it confers some reproductive benefit on its possessor (or on other members of the same species, as in kin selection). Larry Wright criticized Ruse on the grounds that Ruse's view did not recognize a distinction between a trait's function and some accidental benefit (Wright 1972, 513). In response to Wright's criticisms, Ruse (1973a, 278) amended his view to specify that a trait only has a function when *either* that trait was selected for in the past, *or* it is currently undergoing selection, *or* it will undergo selection in the future, for that benefit (also see Ruse 1973b, 184, *fn.* 1). Another way of putting it is that what converts a beneficial effect of a trait into a function is that it makes the trait *prone to be selected* over something else. This modification succeeds, it seems to me, by virtue of incorporating a statistical element, since in order for T to be selected for F over some other trait T^*, there has to be a large enough number of tokens of T, and they have to perform F frequently enough to undergo selection.

Bigelow and Pargetter (1987, 192) also define function in terms of fitness-contributions. However, instead of saying that a trait has a function when it confers some statistically-typical benefit onto the species, they say that a trait has a function when it confers on the organism a *propensity* to contribute to fitness in that organism's natural habitat. I will address their view on function more in the next section. The important point, for now, is that a propensity to contribute to fitness is a sort of disposition that the organism has, one that manifests itself in a probabilistic way. It is a property that the organism possesses even when the trait is not, in fact, contributing to fitness. By the same token, salt is soluble—it is disposed to dissolve in water—even if it is never placed in water. Something can have a disposition even if it never manifests that disposition. So, I take their viewpoint to be very similar to Ruse's in that the function of a trait is defined in terms of its proneness to selection. Bigelow and Pargetter emphasize that their view of function differs from Boorse's on account of the fact that it identifies functions with the propensity that a trait possesses, rather than in terms of the frequency of occasions on which it manifests that propensity (193, *fn.* 9).

Walsh (1996) promotes a view he calls the "relational" theory of function (also see Walsh and Ariew 1996). In his view, we should never say that a trait has a function *simpliciter*. Rather, we should say that it has a function relative to some specified environment or "selective regime," and we should state explicitly what that regime is. As he puts it, "The/a function of a token of a type X with respect to a selective regime R is to m iff X's doing m positively (and significantly) contributes to the average fitness of individuals possessing X with respect to R" (564). I will describe this view in more detail in the next section. The point here is that Walsh and Ariew clearly wish to define the function of a trait in terms of its proneness to selection (relative to some specified environment), and to that extent, it shares a common theme with Ruse's view and with Bigelow and Pargetter's view.

Wouters (1995, 2005, 2013) has developed a view of function that does not fit naturally into the family of views discussed above, but that shares a common commitment to the idea that the function of a trait has to do with its current-day contribution to the survival and reproductive prospects of the organism that has it. He thinks that functions play a role in a distinctive style of explanation, which he calls "viability explanation," and which he thinks philosophers of biology have overlooked. In this style of explanation, a biologist explains a trait by showing how that trait satisfies a need of the organism (the needs generated by its environment and its other traits). For example, suppose we want to know why mammals require a circulatory system, rather than a simple oxygen-diffusion system. The answer is that mammals need a circulatory system because of their large size (1995, 446). Another way he puts the view is by saying that a trait has a function when it positively affects the life chances of an organism *relative* to some real or hypothetical variant (2005, 42). I will come back to this view in the next section. Note that viability explanation is not a type of causal explanation. After all, merely by showing how a trait fulfills a biological need, I am not thereby showing how that trait came to exist. Wouters believes the notion of explanation is broad enough to include non-causal explanations (1995, 437).

4.2 Three Questions for the Fitness-Contribution View

There are at least three more questions that any version of the fitness-contribution view must answer. First, is the notion of a fitness-contribution comparative? That is, when we say that T contributes to fitness, are we implicitly comparing it with an alternative trait (say, T^*)? If so, then how do we specify, in a non-arbitrary way, the set of alternatives? Note that fitness is not always used as a comparative notion. In some contexts we assess the fitness of an organism in terms of the number of expected offspring, regardless of how it compares with others. The point here is that, *if* we interpret fitness as a relational notion and then define function in terms of fitness, we have to specify what the functional trait is being compared to.

It is natural to think that organisms with T are simply being compared to hypothetical organisms without T (as in Canfield's solution, described above). But it is notoriously difficult to specify the relevant counterfactual situation (Frankfurt and Poole 1965; Wimsatt 1972, 2013; Wright 1972). For example, if we want to know the function of the mouse's bile, we would have to imagine what it would be like for there to be a mouse whose gall bladder did not produce bile and how well it would fare. But what else would be true of this mouse? In this counterfactual situation, would this mouse have some other way of breaking down food? What would its digestive system be like? Whether this mouse fares well or poorly in the comparison depends on what exactly we picture this mouse to be like.

One way to solve this problem is through sheer stipulation. We can just relativize the function of a trait to a stipulated set of real or hypothetical variants. For example, we can say that the function of the heart in mammals is to pump blood, relative to a hypothetical mammal that utilizes a simple oxygen-diffusion system (e.g., Wouters 2003, 645). One might worry that this solution raises the threat of undisciplined relativism about function ascriptions. For example, for Wouters, we can say that the reason zebra stripes have the *function* of deterring flies is that a zebra with stripes would be better at deterring flies than, say, a hypothetical monocolored zebra. In that particular comparison, the striped zebra would come out ahead. But in comparison with, say, a hypothetical monocolored zebra that could just spray a massive cloud of bug repellent from pores along its body, the striped zebra would come out behind, so stripes would not have that function. So whether or not it has the function depends on what contrast we have in mind.

On reflection, however, the situation probably is not so bad. Wouters could just respond by saying that, when we assign functions to traits, we do so against a background of largely tacit assumptions about what the relevant alternatives are. If those background assumptions were to change substantially, then some of the functions statements that we now consider to be true would turn out to be false. Functions have a relational and comparative character; it is just that in ordinary discourse we often leave those assumptions unstated.

Another approach, which I think is preferable, is to say that a contribution to fitness need not be comparative (suggested by Buller 1998, 509). For example, a purely physiological analysis of my ability to survive would yield the result that my lungs help me to breathe, and breathing contributes to my survival, even if I am not implicitly comparing my survival prospects to those of a hypothetical individual without lungs. This maneuver would be very useful for the fitness-contribution theorist because it entirely circumvents the need to specify some set of alternatives when we attribute a function to a trait. (Buller himself does not accept the fitness-contribution theory, but a version of the etiological theory, and I will describe his view in some detail in Sect. 6.1.)

There is a second question that the fitness-contribution theorist must answer. What constitutes the *normal environment* for a trait? When we say that a trait contributes to fitness, we do not mean that it contributes to fitness in every environment. The polar bear's thick fur contributes to fitness in the Arctic Circle; if we moved the polar bear to a tropical zoo, that same trait would lower its fitness. So when we say that the trait contributes to fitness, we mean, it contributes to fitness in certain environments. If we define function in terms of fitness, then functions, too, are relative to environments. So how do we specify (again, non-arbitrarily) this set of environments?

Bigelow and Pargetter (1987, 192) relativize the function of a trait to that trait's *natural habitat*. They are content to leave this idea vague. They suggest, however, that they are willing to characterize the natural habitat for a trait historically, as the environment that the species inhabited in the recent past. But note the paradox here. One of the main arguments for the fitness-contribution view is that it dispenses with historical considerations! Bigelow and Pargetter's move would effectively introduce historical considerations into the fitness-contribution theory, which would bring the fitness-contribution view closer in spirit to the selected effects view. What that means is that, if the reason that one is inclined to accept the fitness-contribution view is that it doesn't appeal to history (and one finds that a desirable feature of a theory) then one might be mistaken about what the theory is committed to.

Perhaps instead of introducing the problematic notion of a normal habitat, we can rely on statistical considerations. Suppose we say that the function of trait T is to do F, because T's doing F makes a non-negligible contribution to fitness (considered as a conditional probability, as in Sect. 4.1). Then, perhaps we can just say that the "normal environment" for T (relative to F) is that environment within which T's doing F typically does, in fact, contribute to fitness (as in Garson and Piccinini 2014, who use statistical considerations to define the notion of an "appropriate situation" for a trait's function). The normal environment for my gag reflex is the environment that includes asphyxiation, because that is the environment that the gag reflex is typically in when it makes a non-negligible contribution to fitness. Note that this does not dispense with historical considerations because when we assess the typical environment in which a trait contributes to fitness we have to consider a slice of time that stretches back into the past (as noted by Boorse 2002, 99). As a consequence, the distinction between the selected effects theory and the fitness-contribution theory should not be framed as a distinction between historical and ahistorical theories.

Another possibility for solving the problem of the reference environment is by sheer stipulation. We can simply accept that function statements are always relative to a stipulated environment. This is what Walsh (1996) and Walsh and Ariew (1996) do. They hold that the function of a trait is relative to a selective regime, and this regime must be specified explicitly in order to decide whether any given function attribution is correct. I think their view implies that we should say that relative to the cold, Arctic environment, the function of the polar bears' thick fur is heat-retention, but relative to the tropics, the fur has no function at all (or, perhaps, that it is dysfunctional). Like Wouters' view described above, this raises the threat of a sort of undisciplined relativism about function assignments, but they could avoid this by appealing to the role of background assumptions. This view is attractive because it would neither need to appeal to history nor to the normal environment for a trait.

Here is a third and final question that the fitness-contribution theorist must answer. *What exactly* must a trait contribute to in order for that contribution to count as a function? Writers are typically vague on this score, and use terms such as "fitness," "survival," "survival and reproduction," or even "persistence" (Sarkar 2005, 18; also see Weber 2005, 39, who uses the term "self-reproduction"). But we can take each in turn to suggest that it is not quite what we want.

First, we cannot say that a trait has a function when it contributes to the *survival* (or persistence) of the organism. For some traits, the performance of the function kills the creature (for example, when a female praying mantis eats its partner's head during copulation). Nor can we restrict functions to those traits that contribute to *reproduction*, for then we could not attribute functions to the parts of sterile animals such as mules. If we restrict functions to those that contribute to *survival or reproduction*, meant inclusively, then we encounter problems when it comes to inclusive fitness. A honeybee's sting contributes neither to her survival nor her reproductive ability, but to her mother's reproductive ability. We can use the term "fitness" (incorporating inclusive fitness). That is probably the wisest approach, though it does draw us into somewhat subtle debates about the very meaning of "fitness" (see Rosenberg and McShea 2008, Chap. 3). To give an example of the complexities here, suppose we define fitness as average number of offspring. We still do not capture everything that is relevant to the evolutionary trajectory of the species. For example, we do not capture the evolutionary benefits for those organisms that can modify their own sex ratio (the number of male to female offspring) and thereby maximize their total number of grand-offspring (Griffiths 2009).

4.3 An Assessment

Stepping back, this section will assess the theory more globally. First, supposing that we can answer the three questions in the last section in a satisfactory way, why would one accept the theory? In other words, is there any good argument for the view? Second, how does the view fare with respect to the three desiderata spelled out in Sect. 1.2 (the function/accident distinction, and the explanatory and normative dimensions of functions)?

It seems to me that the chief argument for the fitness-contribution view is simply that it reflects actual biological usage well, and in fact it reflects usage much better than the selected effects view. (At the same time, it imposes principled biological restrictions on what sorts of items can have functions, unlike the causal role view—see next chapter.) The reason it is supposed to be closer to biological usage is that biologists don't appeal to history when they assign functions to traits (at least, in a substantial proportion of cases). Rather, they just appeal to current performance (e.g., Amundson and Lauder 1994; Walsh 1996; Wouters 2003).

What exactly do we mean when we say that biologists "don't appeal to history" when they talk about functions? This is supposed to be a central argument for the theory, so it would be good to spell it out a bit more clearly. There are several ways one might interpret this claim that biologists don't appeal to history. I will describe three ways one might understand this claim, and I will explain why I do not find any of them convincing.

The first is that, when biologists attribute functions to traits, they are not *consciously*, or *explicitly*, thinking about history. This is a kind of sociolinguistic claim and I suppose that it is often true. But that just implies that biologists use the term "function" in a largely unreflective or unanalyzed way. When philosophers purport to give a conceptual analysis of a certain term, like "causation," or "knowledge," they are not necessarily explaining what people consciously or explicitly have in mind when they use the term. They are telling us what people *tacitly* have in mind when they use a term, even if those facts are not readily apparent to them. It is not readily apparent to ordinary people that the concept of knowledge bears some reference to the notion of a justified belief. It is not readily apparent that the notion of causation bears reference to counterfactual states of affairs. But those facts have never been leveraged against of the justified true belief theory of knowledge, or against the difference-making theory of causation. So, the fact that biologists are not consciously thinking about natural selection when they attribute a function to something does not mean that some reference to natural selection cannot enter into our best analysis of what "function" means.

Here is an illustration of that point. Some philosophers of biology have described the fitness-contribution view as an "ahistorical" theory of function. By this, they seem to mean that the theory does not make any reference to history at all. But as we have seen above, this might be a mistake. There are sophisticated articulations of that theory that include a historical element. That historical element was not readily apparent to philosophers when they first articulated the fitness-contribution view. Now it has become more apparent. Maybe when biologists think they are not talking about history, they really are talking about history and they just have not thought the matter through carefully enough to see that.

A second way of understanding this claim that biologists don't appeal to history might be the following. One could argue that the fitness-contribution theory best reflects what biologists *say* about their use of function. For example, Tinbergen (1963) famously distinguished functional from evolutionary questions in biology, and equated the former with current survival value, and not history. Mayr (1961) made a similar distinction. Yet, as I showed in Sect. 3.1, other notable biologists stated explicitly that by "function" they mean a trait's selected effect. So, appealing to biologists' explicit definitions of "function" will yield, at best, pluralism about function, which I endorse.

A third way to understand the claim that biologists don't appeal to history is that, when biologists wish to evaluate competing claims about function, they typically look to some subset of current-day benefits of the trait. In other words, when biologists disagree with each other about what the function of a trait is, they typically look at how the trait in question currently contributes to fitness *now*. When they have discovered how the trait currently contributes to fitness, that settles the debate about function. So, the idea here is that the fitness-contribution theory best reflects actual biological practice, regardless of how it fares as a piece of conceptual analysis.

There is some truth to this last claim. When we want to know the function of eyespots on butterfly wings, we look at what eyespots *currently* do for the butterfly in natural settings (as in Prudic et al. 2015). But this datum is compatible with the selected effects theory. Even when biologists explicitly construe function in terms of selected effects, they *also* look to some subset of current-day benefits to discover the function of a trait. This is because they believe that looking to current day benefits is the best (most efficient, most convenient) way of gathering evidence about what it was selected for (as in Caro et al. 2014). So the fact that biologists routinely look to present-day behavior when resolving debates about function is consistent with both the selected effects and the fitness-contribution views.

A different way of putting this point is to distinguish between *constitutive* relations and *evidential* relations. Why do biologists look to current fitness-contributions in order to resolve debates about function? The answer that the fitness-contribution theorist gives is: current fitness contributions are *constitutive* of what the function of something is. The answer the selected effects theorist gives is: current fitness contributions provide *good evidence* for what the function of something is. Appeal to biological practice alone will not settle the issue for us.

How does the view fare with respect to the three desiderata that were laid out in Chap. 1? The view can certainly handle the distinction between function and accident, by appeal to the appropriate statistical norm. We can say that the function of the nose is to help us breathe, and not hold up glasses, because the former constitutes its statistically-typical contribution to fitness (or something along those lines, noting the difficulties involved in defining "typical").

Can the fitness-contribution view make sense of the explanatory dimension of functions? If by "explanation," we mean *causal* explanation, it cannot. That is, if function ascriptions (sometimes) constitute causal explanations for the present-day existence of a trait, then fitness-contribution views cannot account for this fact. Of course, one can appeal to *past* contributions to fitness to explain the current existence of traits. But the fact that a trait *currently* has a function does not explain the current existence of that trait. It seems to me that a core advantage of the selected effects theory over the fitness-contribution theory is that it captures *this* explanatory role of function ascriptions in a natural way.

Proponents of fitness-contribution views may claim that appealing to fitness is *also* explanatory, but then change the *explanandum* of the function statement. In other words, the selected effects theorist and the fitness contribution theorist can agree that functions are explanatory, but disagree about what functions explain. The selected effects theorist claims that function statements provide causal explanations for the existence of the trait, or the current prevalence of the trait in the population. But perhaps function statements can be understood as explaining how a trait contributes to survival, when it does contribute to survival (Bigelow and Pargetter 1987, 193). Maybe function attributions purport to explain why certain individuals are fitter than others, in terms of their current-day properties (Walsh 1996, 571). So, whether or not fitness-contribution theories are explanatory partly depends on what we take the *explanandum* to be.

Another way to restore explanatory power to fitness-contribution functions is to argue that not all explanations are causal explanations. Perhaps there are legitimate non-causal explanations for the existence of traits. Perhaps we should say that, on the fitness-contribution account, functions are explanatory, but they provide a non-causal sort of explanation. This would seem to allow fitness-contribution functions to satisfy the explanatory desiderata.

For example, Hardcastle (1999, 30) states that there are two ways to explain why an organism has a trait. The first is to explain it in terms of its causal history, e.g., in terms of what it was selected for. The second is to explain it in terms of an advantage that it currently gives to that organism, irrespective of causation. Maybe the fitness-contribution view explains the presence of a trait in this latter, non-causal sense. Wouters (1995) also has a non-causal notion of explanation in mind when he says that functions explain why an organism has a trait. They explain, he thinks, by way of viability explanation, which is a kind of non-causal explanation. Similarly, Craver (2013, 155) ventures to suggest that, in addition to what he calls "etiological" and "constitutive" explanations, there is a third sort of explanation called a "contextual" explanation. He states that sometimes, when we ask why a trait is there, we just want to know how it relates to the other components of a mechanism in the style of a contextual explanation. Both of these maneuvers—to change the *explanandum* or to argue that function statements are non-causal explanations— seem strained to me, but I realize that not everyone shares my intuitions.

Finally, are functions normative on this account? Can we explain what it is for a trait to possess a function that it cannot perform? On some explications of the view, it is hard to see how this could be done. For example, Bigelow and Pargetter (1987) define the function of a trait in terms of dispositions. They say a trait has a function when it confers a survival-enhancing disposition upon the organism. Yet to say that a trait malfunctions, for example, that a person's eyes cannot see because that person is congenitally blind, is to say that it *cannot* confer such a disposition onto the organism. So it would not, for that reason, possess a function at all.

It would seem that the best way to account for the normativity of function would be to appeal, again, to a statistical norm. Very roughly, a trait token *malfunctions* when it cannot do what tokens of that trait typically do that contributes to fitness (along the lines of Walsh 1996, 568). But I find this definition of malfunction counterintuitive, for the following reason. Suppose there is a new, beneficial mutation in a population, say, a new sort of pesticide-resistance in a flour beetle. Initially, only one member of the population has the new trait. Those that do not have it are not intuitively malfunctioning on that account, consistently with Walsh's view. Now, suppose that trait begins to spread through the population by natural selection. *Before* it reaches, say, 50 % of the population, those that do not have it are *not* malfunctioning. *After* it reaches over 50 % of the population, those that do not have it *are* malfunctioning. So, whether or not a trait is malfunctioning depends on what is going on in other members of the population. But it seems to me that when we say that something is malfunctioning, what we mean is that there is something like an inner, constitutional breakdown or defect that prevents it from performing its function. Nothing like that is happening among the sub-population of flour beetles that do not have the new mutation. It is certainly *disadvantageous*, but there is a difference, it seems to me, between saying that a trait is disadvantageous and saying it is malfunctioning.

Moreover, as Neander (1991) points out, this statistical approach to normativity raises a new problem, that of pandemic diseases. The problem is simple. It is possible, both logically and physically, for most instances of a trait to malfunction. For example, the Cascades frog in Oregon has been widely afflicted, on a population-level, with an immune-system malfunction that might have resulted from an increase of ultraviolet B-radiation (Sarkar 1996). This would be an example of a population-wide malfunction. Walsh's suggestion—that a trait token is malfunctional if it cannot do what most other tokens can do—would not allow us to say, of any given Cascades frog's immune system, that it is malfunctioning. Interestingly, there is quite a substantial bit of literature now on precisely this question of whether statistical approaches to function can make sense of the notion of dysfunction (Kingma 2010; Hausman 2011; Kraemer 2013; Garson and Piccinini 2014; Boorse 2014).

I think the fitness-contribution view can avoid this problem of pandemic disease by emphasizing the historical dimension of functions. The function of trait type *T* is to do *F* because that is what most of its instances have done that contributed to fitness within a certain era of time including the present and stretching back into the past (Boorse 2002; Garson and Piccinini 2014). This would avoid the problem, but note that historical considerations are doing a lot of work for the theorist, which undermines some of its advantage over the selected effects view.

References

Amundson, R., & Lauder, G. V. (1994). Function without purpose: The uses of causal role function in evolutionary biology. *Biology and Philosophy, 9*, 443–469.

Bechtel, W. (1986). Teleological function analyses and the hierarchical organization of nature. In N. Rescher (Ed.), *Current issues in teleology* (pp. 26–48). Lanham, MD: University Press of America.

Bigelow, J., & Pargetter, R. (1987). Functions. *Journal of Philosophy, 84*, 181–196.

Boorse, C. (1976). Wright on functions. *Philosophical Review, 85*, 70–86.

Boorse, C. (1977). Health as a theoretical concept. *Philosophy of Science, 44*, 542–573.

Boorse, C. (2002). A rebuttal on functions. In A. Ariew, R. Cummins, & M. Perlman (Eds.), *Functions: New essays in the philosophy of psychology and biology* (pp. 63–112). Oxford: Oxford University Press.

Boorse, C. (2014). A second rebuttal on health. *Journal of Medicine and Philosophy, 39*, 683–724.

Buller, D. J. (1998). Etiological theories of function: A geographical survey. *Biology and Philosophy, 13*, 505–527.

Canfield, J. (1964). Teleological explanation in biology. *British Journal for the Philosophy of Science, 14*, 285–295.

Canfield, J. (1966). Introduction. In J. Canfield (Ed.), *Purpose in nature* (pp. 1–7). Englewood Cliffs, NJ: Prentice-Hall.

Caro, T., et al. (2014). The function of zebra stripes. *Nature Communications, 5*, 3535.

Craver, C. (2013). Functions and mechanisms: A perspectivalist view. In P. Huneman (Ed.), *Function: Selection and mechanisms* (pp. 133–158). Dordrecht: Springer.

Frankfurt, H. G., & Poole, B. (1965). Functional analyses in biology. *British Journal for the Philosophy of Science, 17*, 69–72.

Garson, J., & Piccinini, G. (2014). Functions must be performed at appropriate rates in appropriate situations. *British Journal for the Philosophy of Science, 65*, 1–20.

Griffiths, P. E. (2009). In what sense does 'nothing make sense except in the light of evolution'? *Acta Biotheoretica, 57*, 11–32.

Hardcastle, V. G. (1999). Understanding functions: A pragmatic approach. In V. G. Hardcastle (Ed.), *Where biology meets psychology: Philosophical essays* (pp. 27–43). Cambridge, MA: MIT Press.

Hausman, D. (2011). Is an overdose of paracetamol bad for one's health? *British Journal for the Philosophy of Science, 62*, 657–668.

Horan, B. (1989). Functional explanations in sociobiology. *Biology and Philosophy, 4*, 131–158.

Kingma, E. (2010). Paracetamol, poison, and polio: Why Boorse's account of function fails to distinguish health and disease. *British Journal for the Philosophy of Science, 61*, 241–264.

Kraemer, D. M. (2013). Statistical theories of functions and the problem of epidemic disease. *Biology and Philosophy, 28*, 423–438.

Lehman, H. (1965). Functional explanation in biology. *Philosophy of Science, 32*, 1–20.

Lewens, T. (2004). *Organisms and artifacts: Design in nature and elsewhere*. Cambridge, MA: MIT Press.

Mayr, E. (1961). Cause and effect in biology. *Science, 134*, 1501–1506.

Millikan, R. G. (1984). *Language, thought, and other biological categories*. Cambridge, MA: MIT Press.

Neander, K. (1991). Functions as selected effects: The conceptual analyst's defense. *Philosophy of Science, 58*, 168–184.

Prudic, K. L., et al. (2015). Eyespots deflect predator attack increasing fitness and promoting the evolution of phenotypic plasticity. *Proceedings of the Royal Society B, 282*, 201415.

Rosenberg, A., & McShea, D. W. (2008). *Philosophy of biology: A contemporary introduction*. New York: Routledge.

Ruse, M. E. (1971). Functional statements in biology. *Philosophy of Science, 38*, 87–95.

Ruse, M. E. (1973a). A reply to Wright's analysis of functional statements. *Philosophy of Science, 40*, 277–280.

Ruse, M. E. (1973b). *The philosophy of biology*. Atlantic Highlands, NJ: Humanities Press.

Sarkar, S. (1996). Ecological theory and anuran declines. *BioScience, 46*, 199–207.

Sarkar, S. (2005). *Molecular models of life*. Cambridge, MA: MIT Press.

Tinbergen, N. (1963). On aims and methods of ethology. *Zeitschrift für Tierpsychologie, 20*, 410–433.

Walsh, D. M. (1996). Fitness and function. *British Journal for the Philosophy of Science, 47*, 553–574.

Walsh, D. M., & Ariew, A. (1996). A taxonomy of functions. *Canadian Journal of Philosophy, 26*, 493–514.

Weber, M. (2005). *Philosophy of experimental biology*. Cambridge: Cambridge University Press.

Wimsatt, W. C. (1972). Teleology and the logical structure of function statements. *Studies in the History and Philosophy of Science, 3*, 1–80.

Wimsatt, W. C. (2013). Evolution and the stability of functional architectures. In P. Huneman (Ed.), *Function: Selection and mechanisms* (pp. 19–41). Dordrecht: Springer.

Wouters, A. G. (1995). Viability explanation. *Biology and Philosophy, 10*, 435–457.

Wouters, A. (2003). Four notions of biological function. *Studies in the History and Philosophy of Biological and Biomedical Sciences, 34*, 633–668.

Wouters, A. G. (2005). The functional perspective in organismic biology. In T. A. C. Reydon & L. Hemerik (Eds.), *Current themes in theoretical biology* (pp. 33–69). Dordrecht: Springer.

Wouters, A. G. (2013). Biology's functional perspective: Roles, advantage, and organization. In K. Kampourakis (Ed.), *The philosophy of biology: A companion for educators* (pp. 455–486). Dordrecht: Springer.

Wright, L. (1972). A comment on Ruse's analysis of function statements. *Philosophy of Science, 39*, 512–514.

Wright, L. (1973). Functions. *Philosophical Review, 82*, 139–168.

Chapter 5
Functions and Causal Roles

Abstract This chapter is about the causal role theory of function. According to this view, roughly, a function of a part of a system consists in its contribution to some system-level effect, which effect has been picked out as especially interesting by a group of researchers. I will discuss Robert Cummins' original formulation of the view, and then present a more sophisticated variation, the mechanistic causal role theory, due to Carl Craver and Paul Sheldon Davies. I then discuss the classic problem of overbreadth, namely, that it seems to attribute functions too liberally. I distinguish two different versions of this problem: the problem of non-functional traits and the problem of dysfunctional traits. I provide a critical assessment of the ways causal role theorists have tried to solve these problems. Many philosophers of biology today have accepted a pluralistic stance, according to which both the selected effects theory, and the causal role theory, capture important elements of biological usage. I distinguish two forms of pluralism. The first (and most popular), between-discipline pluralism, holds that the selected effects theory mainly captures the way evolutionary biologists use the term "function" and the causal role theory mainly captures the way "function" is used in other disciplines. I object to this division of labor and recommend a new form of pluralism, within-discipline pluralism, which emphasizes the co-existence of function concepts within any given discipline.

Keywords Causal role functions · Cummins functions · Mechanistic causal role theory · Function and mechanism · Function pluralism

5.1 From Causal Roles to Mechanisms

At least sometimes, when biologists attribute a function to an item, they seem to be doing little more than showing how the item promotes some capacity of a larger system. The kidney is made up of a large number of tubes called nephrons. Each contains a glomerulus, which functions to extract water out of the bloodstream, a tubule, which functions to remove nutrients from the water, and a collecting duct, which functions to convey the water to the bladder. Collectively, these nephrons allow the kidney to filter blood and remove excess water from the body.

When I say that the glomerulus "functions to" filter water out of the bloodstream, it seems that I am just describing the part it plays in a complex system or mechanism. Moreover, attributing a function *qua* role to an item seems to be a different sort of explanatory project, or focus, than asking why the item is there in the system. On the face of it, one could figure out the role of the glomerulus in helping the kidney without knowing anything about how it evolved. (After all, it could have evolved for one thing, and then later started becoming useful for something else.) In some cases it is unclear what sort of function a biologist is talking about when he or she says, "the function of the glomerulus is to filter water from the bloodstream." Is that biologist trying to tell us why we have glomeruli? Or is that biologist trying to tell us something about the role they play in a complex system?

So, there are two questions here: can we explicate this "causal role" sense of function in a satisfying way, and how does this causal role sense of function relate to the sense outlined in Chap. 3, where functions are thought of in a causal-explanatory, why-is-it-there sense? If there is only *one* notion of function in biology, then these two theories of function, the selected effects theory and the causal role theory, should be seen as competitors. If there are at least *two* notions of function in biology, then both theories can be correct, and we should accept a pluralist account of function in biology. But if we accept pluralism, then can we say anything intelligent about when, or in which contexts, biologists use the one sense and when, or in which contexts, biologists use the other? So, although this chapter is presumptively about causal roles, it will inevitably tackle broader issues about pluralism.

Cummins (1975, 1983) is the chief proponent of this causal role theory of function, although the theory has been developed, and refined, by others, such as Prior (1985); Amundson and Lauder (1994); Hardcastle (1999, 2002); Davies (2001); Craver (2001, 2013), Šustar (2007), and Šustar and Brzović (2014). In this section I will describe Cummins' original version of the theory, and then discuss Craver's fairly sophisticated development of the idea.

Cummins did not think function statements are explanatory, in the sense of giving a causal explanation for a trait. I discussed his criticisms in Sect. 3.3. He was also skeptical about the fitness-contribution view, which identifies functions with fitness-enhancing benefits. He thought that biologists sometimes attribute functions to traits regardless of whether or not they benefit the species: "Flight is a capacity that cries out for explanation in terms of anatomical functions regardless of its contribution to the capacity to maintain the species" (1975, 756).

So what sort of explanation are we giving when we say, "the function of the heart is to pump blood?" We are not trying to explain the existence of the heart. Instead, we are trying to explain blood circulation. But we are offering a very special sort of explanation. Cummins contrasted functional explanations with another sort of explanation, which he called the "subsumption" strategy. Suppose I want to explain why an item *i* has a disposition *d*. In some cases, I explain why *i* has *d* by "subsuming" *i* under some basic physical laws and some boundary conditions.

For example, suppose I want to explain why salt is soluble. That explanation falls out of the laws of chemistry and some basic boundary conditions. Salt is made up of sodium and chloride, and when you put it in water, the positive charge of the water attracts the chloride, and the negative charge of the water attracts the sodium, and they come apart.

Other dispositions do not as neatly fall out of basic physical laws, such as how the kidney filters blood, or why people get startled when they see snakes, or why sub-prime mortgages created a housing bubble that devastated the US economy. These phenomena call out for a different sort of explanation. These are "functional" explanations. They have the following structure. First, we pick out some system property or capacity that we are interested in (e.g., the filtration of the blood by the kidney). Then we decompose the system into a number of sub-capacities (which he called its "analyzing capacities"), e.g., the capacity to extract water, to remove nutrients, and to convey the water to the bladder. We then show how those sub-capacities are organized so that they yield the capacity as a whole (namely, in a simple series). The function of any given item is simply the role that it performs in the context of a functional explanation. The glomerulus has the function of extracting water from the blood because that is the role it plays in a functional account of how the kidney filters the water.

How do we know which explanatory strategy to use on a given occasion? Cummins (1975, 764) said that there were three marks, or indicators, for when we should use functional analysis (rather than the subsumption strategy). These are rough indicators and are not intended to be very precise criteria. Functional analysis is appropriate when, first, the sub-capacities are less sophisticated than the analyzed capacity, second, the sub-capacities are different in type than the analyzed capacity, and third, the organization of the sub-capacities is relatively sophisticated.

Here is a simple example where it would not be reasonable to use functional analysis. Suppose I want to explain the weight of a pile of rocks, in terms of the weights of its component rocks. Clearly, the sub-capacities (the weights of the individual rocks) are not less sophisticated, or different in type, than the capacity we are trying to account for (the weight of the pile). Moreover, the relevant organization is not very sophisticated. The rocks are just piled up haphazardly on top of each other. Wimsatt's (1986) interesting account of "organization" versus "aggregativity" is useful here; both Craver (2001, 59) and Davies (2001, 82) refer to Wimsatt's account to help us understand functional analysis.

So far, so good. The function of an item is just its role in a functional explanation of some system capacity. But here is a problem. Any given item, such as the heart, plays a role in many different systems, and hence it has a large number of functions. The heart plays a role in the circulatory system, by pumping. The heart plays a role in the diagnostic system, by making throbbing sounds that one can listen to through a stethoscope. The heart also plays a role in polygraph tests, by supposedly indicating when somebody is telling a lie by causing a spike in blood pressure. So, Cummins' theory of function seems to imply that the heart has many, many different functions: beating, making throbbing sounds, and causing spikes in blood pressure. But it would be strange for a physiology textbook to list all of these as the heart's functions. So, Cummins' theory is overly liberal in the way that it attributes functions to things. (I will develop this critique in the next section. There are actually two different liberality problems.)

This problem led Cummins to emphasize the idea that functions have to do with our own interests and perspectives. When we attribute a function to something, we are always "speaking against the background of an analytical explanation of some capacity..." (762). Functions are always relative to some capacity that we happen to be interested in. Biologists have been interested in the mechanism of blood circulation for a long time. But they have not been particularly interested in the mechanism by which beating sounds can be heard through a stethoscope. That is why the statement, "the function of the heart is to pump blood" seems natural and correct, and the statement, "the function of the heart is to make beating sounds," sounds odd or wrong. But if we had a very different set of interests, it wouldn't sound wrong at all. Suppose there were some intelligent aliens that came to earth and studied the human body. They might attribute functions to it in ways that would seem weird to us, but that would be entirely correct from their perspective. Other theorists such as Hardcastle (1999, 2002) and Craver (2001, 2013) also emphasize the epistemic and pragmatic dimensions of functions.

Carl Craver's work (2001, 2013) represents a significant development of Cummins' view, though Paul Davies (2001) also made important contributions along similar lines, which I will discuss in the following section. Craver's goal was to re-state Cummins' theory of function in the framework of mechanistic explanation. Craver complains that Cummins' discussion of function is overly vague on several crucial points: for example, on the nature of "organization," "complexity," "capacities," and so on. He thinks the mechanistic framework can help make Cummins' approach more precise. So I will say something about mechanistic explanation, and then I will say how Craver re-casts Cummins' view in that framework.

Since the 1990s, philosophers of science have been systematically articulating the concept of mechanism and identifying the role it plays in the life sciences. Clearly, one important goal of the life sciences is to discover the mechanisms for various phenomena. What is the mechanism for mitosis? What is the mechanism by which mosquitoes infect their hosts with malaria? What is the mechanism of memory? The notion of mechanisms, and the corresponding concept of mechanistic explanation, has received a lot of attention (Bechtel and Richardson 1993; Glennan 1996; Machamer et al. 2000; Darden 2006; Craver and Darden 2013; see Wimsatt 1976 for an important precursor.) Sarkar (1998) describes some of the features of mechanistic explanation under the label of "strong hierarchical reduction."

The basic idea of mechanistic explanation is fairly straightforward. Any mechanism is defined by a certain *phenomenon*. That is, mechanisms are always mechanisms *for* something or another. If we are seeking the mechanism of mitosis, then mitosis is the mechanism's phenomenon. Selecting the phenomenon is the necessary starting point for undertaking a mechanistic explanation. ("How many mechanisms are in the human body?" is not a well-defined question.)

We then decompose the system into various *components*: the physical parts of the mechanism and their associated activities. We then identify the crucial features of *organization* that explain how the components give rise to the phenomenon. These are the spatial and temporal relationships between the components. Finally, mechanisms have hierarchical and serial relationships as well. The fact that mechanisms are often hierarchically nested gives rise to a project called "multi-level mechanistic modeling." Darden (2006, 280) gives a nice synopsis of the elements of a mechanistic explanation.

Now we can begin transposing various elements of Cummins' theory into this newer framework of mechanistic explanation. Where Cummins talked about a "system," we can talk about a "mechanism." When Cummins spoke of the "analyzing capacities" of a system, we can talk about the "components" of the mechanism (parts and activities). Where Cummins talked about the "organization" of the system, we can talk about the spatial and temporal relationships between the parts and the hierarchical embedding of mechanisms. Finally, since the way that we individuate mechanisms is partly dependent on our research perspective, then functions are still relative to our research purposes, just as Cummins believed they were.

This mechanistic causal role theory of function is not merely a matter of shifting terms around or using a new vocabulary. It represents a significant theoretical advance. Here is one key advantage. In Cummins' framework, when we decompose our system into a number of different analyzing capacities, these capacities may or may not correspond to actual physical components. For example, in cognitive science (which is one domain that Cummins is interested in) we sometimes decompose or analyze a given psychological capacity, such as the capacity to multiply two numbers, into various sub-capacities, even when we do not know whether or not those sub-capacities correspond neatly to any physical parts of the brain. Cummins insisted that functional analysis is different from what he called componential analysis. Only the latter is overtly concerned with the identification of physical components of a system (Cummins 1983, 29). Cummins realized that functional analyses are loosely constrained by facts about the physical components of the system. Cummins agreed, as Davies (2001, 26) emphasized, that functional analysis must bottom out in actual physical components of the system. In other words, at the lowest level of functional analysis, the sub-capacities must correspond to actual physical components of a system (Cummins 1983, 31). But, at all intermediate levels of functional analysis, the correspondence between the analyzing capacities and the physical components of the system can be, as Cummins put it, "very indirect" (29). In contrast, in the mechanistic framework, it is very clear that we are always talking about actual physical components of a system at every level of analysis (Craver 2001, 57; also see Piccinini and Craver 2011 for discussion).

5.2 Reining in Causal Roles

There are two major problems that have haunted the causal role theory. Sometimes these two problems are lumped together by commentators as the problem of "overbreadth," but they should be distinguished. I will call the first, the problem of non-functional traits, and the second, the problem of dysfunctional traits. They both have to do with the fact that the causal role theory lets us assign functions to inappropriate things.

The first problem is that the causal role theory lets us assign functions to items that are intuitively *non-functional* (that is, that do not have functions at all). It seems implausible to say that the function of the heart is to make beating sounds, even though, on Cummins' account, there are some contexts in which that statement is correct (see Matthen 1988, 15; Millikan 1989, 294; Kitcher 1993, 390). Many philosophers who have thought deeply about functions have used the beating sounds of the heart as the paradigm of an accident, not a function. It would be nice for a theory of function to yield that as a consequence.

I am not denying that there is *some* sense of function in which we can talk about heart sounds as a "function" of the heart. This is the sense that is often marked linguistically by the expression "functions as." I noted this alternative sense briefly in Sect. 4.1. The heart may *function as* a diagnostic device by making heart sounds. But I still maintain that making sounds is not one of the heart's functions, in the sense of "biological function" that I am interested in. Similarly, in mathematical contexts, one might say that x "is a function of" y. But this is not the sense of function I am interested in here. If someone protested that, "it sounds right to me to say that a function of the heart is to make sounds," then I would suspect that they are using the term "function" with a different meaning than the one I am pursuing.

We have to be very cautious here in stating what exactly the problem is. The problem is not that the causal role theory cannot distinguish function and accident. As I noted above (Sect. 1.2), this is a very traditional desideratum for a theory of function. It would be unwise to ditch it completely. Fortunately, causal role theorists *can* distinguish between function and accident. They just make the distinction relative to some explanatory interest or perspective. If a group of researchers is interested in blood circulation, then, relative to that standpoint, pumping blood will emerge as a function, and making sounds as an accident. The problem of non-functional traits, rather, is that the causal role theory applies the distinction in ways that are counterintuitive. I do not think that there are any contexts in which it is correct to say that the function of the heart is to make sounds. I think the function of the heart is just to pump blood, regardless of one's goals and interests. (Of course, it has other functions too, such as maintaining an appropriate ratio of carbon dioxide and oxygen in the body. Things can have many functions. But I do not believe that making sounds is even one among these functions. I think it is a paradigm case of an accidental benefit.)

Millikan raises a similar example. She points out that, according to Cummins' theory, there are some contexts in which it would be correct to say that the function of clouds is to promote vegetation growth, by causing rain. Interestingly, Millikan (1989, 294) never actually raised the vegetation example as a *criticism* of the causal role theory—a point that she emphasizes in Millikan (2002, 114, *fn.* 1). She raised the point in order to illustrate a difference between Cummins' analysis and her own. But I think it stands as a good criticism of Cummins' view, too. (Again, I think there is *some* sense of function in which we can say, "vegetation growth is a function of clouds." But this sense of function is different from the biological sense of function.)

There is another problem that is not often carefully distinguished from the first, though Hardcastle (1999, 36) is an exception. I think it is much more serious. Cummins' account would let us assign functions to things that are, intuitively, *dysfunctional*. In other words, it is one thing to note that the theory allows one to say, "the heart's function is to make diagnostic sounds." It is another to note that the theory allows one to say, "the function of myelin degeneration is to cause paralysis." Even Cummins (1975, 752) acknowledges that the latter is counterintuitive. There is something ironic about his acknowledgement because, if I read him correctly, he reprimands Ernest Nagel on just this score (see his *fn.* 8) but his own view yields the same conclusion.

Again, we have to be careful here. I am not claiming that the causal role theory cannot make some distinction between function and malfunction. I am not entirely sure whether or not it can. Some people seem to think that it can (Godfrey-Smith 1993, 200; Craver 2001, 72), and some seem to think it cannot (Neander 1991, 181; Davies 2001, 29; Preston 1998, 224). The straightforward argument for why the causal role view cannot explain malfunctioning is that, on Cummins' theory, functions are dispositions. To say that something has the function F is to say that it has the disposition to do F under such-and-such conditions (Cummins 1975, 757). So if a person's heart is damaged and it cannot pump blood, then it loses the disposition to F, and hence it loses the function. So if one wants the causal role theory to provide a notion of malfunction, one must build it in explicitly. The most obvious way to do that is to relativize functions to trait types. That is, one could hold that, in order for something to have a function, it must be the case that at least some tokens of its type can do F under the right circumstances. (I thank Gualtiero Piccinini for emphasizing this point to me.) All of that would have to be worked out in some detail. I gave some indications for how such a theory might go at the end of Sect. 4.3, where I talked about how the fitness-contribution view might handle dysfunction. At any rate, that discussion is not relevant to the point I am making here. My point is somewhat different. Cummins' view implies that in some contexts it would be correct to say that the function of myelin degeneration is to promote paralysis. That sounds wrong, and he admits that it sounds wrong.

So there are two problems here, namely, the causal role view allows us to assign functions to traits that are nonfunctional, and it allows us to assign functions to traits that are dysfunctional. These are two special cases of the more general problem of overbreadth. Note, moreover, that these are not bizarre science-fiction examples like swampmen or cloner-machines. These are paradigm sorts of scientific cases that we should expect a theory of function to yield the right solution to.

Suppose we take these problems of overbreadth seriously. Then there are two natural lines of thought for reining in causal roles, that is, for developing the causal role view in a systematic way so that it avoids both sorts of problems. These are due to Valerie Hardcastle and Paul Sheldon Davies. I think both come short, but for different reasons. Hardcastle's view seems to solve the problem of dysfunctional traits, but not the problem of non-functional traits. Davies' view seems to solve the problem of non-functional traits, but not the problem of dysfunctional traits. Maybe the two solutions could be somehow combined to yield a compelling solution to both problems. There are other possible solutions, of course. One could try to incorporate stronger evolutionary considerations into Cummins view along the lines of Millikan (2002) or Šustar (2007). I take it that this would be antithetical to the spirit of the causal role view, which seeks to make sense of function ascriptions across a wide range of disciplines, including biology, psychology, and the social sciences.

The first is due to Hardcastle in the context of what she calls the "pragmatic" view of functions, and which I see as a version of the causal role theory. In her view, the difference between the function of a trait and a mere effect is not just a matter of the idiosyncratic goals of this or that investigator. Rather, each specialized field of scientific inquiry gets to decide, as it were, which general *sorts* of effects count as an item's functions: "which parameters are chosen as important in assigning functions depends on the domain doing the abstracting" (Hardcastle 1999, 38). Interestingly, the philosopher of science Hempel (1965, 321–322) also suggested a similar discipline-relativity in function ascriptions. He said that the function of an item has something to do with its contribution to the "proper working order" of the system in which it is contained. He realized that the notion of "proper working order" was vague, but he thought that different disciplines have the right to define "proper working order" differently.

For example, the goal of medicine is to promote health. So, from the standpoint of medicine, the function of the heart is to pump blood, *rather than* to cause heart disease, because the former contributes to health. A similar point can be raised about psychology and psychiatry, where function ascriptions are governed by the disciplinary norm of promoting psychological well-being. Over-activity of the amygdala can promote panic attacks but that is not its function. One of its functions is to help regulate our emotional responses in ways that are proportional to external threats.

The same can be said of ecology and conservation biology. A destructive invasive species does not have an ecosystem function because its behavior is contrary to the goals and interests of conservation biology as a whole. This move, I think, solves the problem of dysfunction fairly well. Even though medical researchers are certainly interested in how myelin degeneration causes paralysis, that does not make paralysis the function of myelin degeneration because, as far as medicine is concerned, that is not the right kind of effect to count as a function.

This is a reasonable start to a solution to the problem of overbreadth, but I think there are two problems. First, Hardcastle's pragmatic view would seem to imply that as one moves across disciplines, one would see systematic and predictable shifts in the way those disciplines assign functions to one and the same trait. For example, the purpose of pest toxicology is to kill pests, so if Hardcastle is right, one would expect pest toxicologists to have a completely different construal of the functional organization of the pest's body. They would be licensed to say that the function of the rat's stomach is to distribute poison to the bloodstream. But they do not say things like that (as evidenced by articles in pest toxicology journals—see Garson 2013, 331 for examples), and at any rate, it would seem counterintuitive to do so. Even pest toxicologists would say that the function of the rat's stomach is to digest food. So I do not believe that one sees the sorts of systematic shifts in function assignments that one would expect on the pragmatic view.

Second, even if the pragmatic view solves the dysfunction problem it does not seem to solve the non-function problem. Ecologists and agricultural engineers are certainly interested in promoting vegetation growth, so the pragmatic view would seem to allow them to say, correctly, that the function of clouds is to promote vegetation growth. But, to my knowledge, they do not say things like that. (Again, they may say things like, "vegetation growth is a function of clouds," but this is not the biological sense of function.)

Davies (2001) attempts to resolve the *first* problem of overbreadth (namely, that the causal role view allows us to assign functions to items that are intuitively *non-functional*) by emphasizing that, in order for the components of a system to have functions, the system must have the right sort of complexity. As I noted above (in the last section), a pile of rocks does not have the right sort of complexity to warrant the ascription of functions. So I think it is fair to see Davies as attempting to systematically elaborate a point that was implicit or only gestured at in Cummins' work, when Cummins set out his three rules for the appropriateness of functional analysis (1975, 764).

Specifically, Davies says that functions should only be assigned to the components of *hierarchically organized systems* (2001, 82). For example, he does not think the causal role theorist has to say that the function of clouds is to promote vegetation growth. He thinks that in this case, the system in question probably does not have the right sort of hierarchical organization to license such an implausible ascription (94).

Although I believe this approach may go some ways toward resolving the first version of the problem of overbreadth (the problem of non-functional traits), I doubt it resolves the second (the problem of dysfunctional traits). Even when we restrict functions to hierarchically organized systems (when this constraint is plausibly interpreted) we could still say that the function of myelin degeneration is to promote paralysis. I would assume, without having undertaken a thorough analysis, that the mechanism that achieves this effect has the right sort of hierarchy and complexity. If paralysis is not a good example, then I suppose there must be other diseases the progression of which exhibits the right sort of hierarchy. Creutzfeldt-Jakob disease, which is a prion-related disease, involves a fairly sophisticated inner mechanism (Aguzzi et al. 2008). It involves the propagation of a misfolded state (prions cause normally folded proteins to misfold), and the aggregation of those misfolded proteins into plaques. I do not see how that would be any less complex or hierarchical than the way that heart pumping promotes blood circulation.

5.3   Varieties of Function Pluralism

In bringing this section to a close, I will make some remarks about pluralism. Garson (Submitted for publication) provides a much more detailed exposition. The reason that I raise the problem of pluralism here (rather than in the context of the fitness-contribution theory) is that the most popular form of pluralism in the functions literature maintains that *both* the selected effects view *and* the causal role view capture important features of real biological usage. In other words, when biologists assign functions to items, in some cases they purport to (causally) explain why that item is there. The selected effects theory appears to be the most well-developed philosophical account of this strand of usage (though see Sects. 6.1 and 6.2 for some alternatives). In other cases, biologists merely purport to describe how it contributes to some capacity of a containing system. The causal role theory seems to be the most sophisticated philosophical account of this strand of usage (perhaps with the further restriction that the item in question contribute to fitness or persistence). I think this is a reasonable and conciliatory stance on the topic, despite the fact that both views still have outstanding problems. Authors who accept this sort of pluralism include Millikan (1989, 2002), Neander (1991), Godfrey-Smith (1993), Mitchell (1993), Amundson and Lauder (1994), Allen and Bekoff (1995), Preston (1998), Griffiths (2006), Maclaurin and Sterelny (2008), Garson (2011), Bouchard (2013), and Brandon (2013).

Let's suppose that this sort of pluralism is correct. Can we say anything interesting about when, or in which contexts, biologists appeal to the causal role theory, and when, or in which contexts, they appeal to the selected effects theory? One possibility is that we cannot give any *general* rules for deciding when biologists use the one notion of function and when biologists use the other. The only way we can decide is by undertaking a case-by-case textual analysis.

One sort of pluralism seems to have gained something of a foothold in the functions literature. I call it "between-discipline pluralism." The idea is that different theories of function are appropriate for different scientific disciplines. In other words, the idea is that one can, very roughly, group different sub-disciplines of biology, and perhaps some areas of psychology as well, by the notion of function its practitioners appeal to. These sorts of pluralists also tend to marginalize the significance of the selected effects view to some limited areas of evolutionary biology. Neander (Submitted for publication, Chap. 4) refers to this sort of pluralism as "popular pluralism" and raises an objection against it. Her view is that selected effects functions are ubiquitous throughout biology because they are implicated in the way that biologists make generalizations about traits. When biologists make generalizations about traits they often describe how the trait would behave if it were "functioning normally." But the selected effects theory gives us perhaps the best way of understanding this idea of normal function. So, when physiologists appeal to the idea of normal function they are probably implicitly committed to the selected effects theory of function. I think she is making an important point here, namely, even when biologists do not explicitly appeal to selection when they attribute functions to traits, they might do so implicitly. I will give some examples of this sort of pluralism to illustrate the style of reasoning.

Godfrey-Smith (1993) endorses this form of pluralism. As he puts it:

> Once a modified version of [the selected effect] theory is in place, the explanatory role of many function statements in fields like behavioral ecology is clear. But there remain entire realms of functional discourse, in fields such as biochemistry, developmental biology, and much of the neurosciences, which are hard to fit into this mold, as functional claims in these fields often appear to make no reference to evolution or selection. (200)

He then goes on to argue that some version of the causal role theory is most appropriate for those latter disciplines.

Amundson and Lauder (1994) make a similar claim regarding anatomy and morphology (anatomy mainly treats of internal structures while morphology treats both internal and external structures). Amundson and Lauder use "function," I believe, in the sense that it assumes when it is distinguished from "form," as in the "form/function" distinction. Here, "form" is more closely aligned with physical structure, and "function" is more closely aligned with activity or performance. For example, if one is investigating the "form and function" of the patella bone (kneecap) one might describe its shape and connectivity ("form") and the basic activity ("function") that it permits (knee extension). It seems evident that "function" here is not, at least explicitly, a historical concept, but should be analyzed along the lines laid out by Bock and von Wahlert (1965), where the function of a trait includes "all physical and chemical properties arising from its form" (274). (Amundson and Lauder cite this paper approvingly.) It also seems that the "function" of a trait, in this minimal sense, can be determined through straightforward experimental manipulation.

They claim that, *as far as morphology is concerned*, the most relevant sense of function is the causal role sense. They write, "the field of biology called functional anatomy or functional morphology explicitly rejects the exclusive use of the [selected effects] concept of function. To be sure, there are other biological fields in which the SE concept is the common one—ethology is an example" (446). Hence they endorse, "the usefulness of different concepts in different areas of research" (ibid.). So, while Amundson and Lauder accept pluralism, they suggest that selected effects functions will not find a home in evolutionary morphology.

Griffiths (2006) endorses their claim, and seems to go further by suggesting that the selected effects theory has little role in those disciplines that constitute "experimental biology" very generally (also see Weber 2005, 38). He says, "unless anatomy, physiology, molecular biology, developmental biology, and so forth turn their attention to specifically evolutionary questions, they investigate function in the causal sense" (3). Other theorists have argued that the selected effects theory of function has little place in ecology, either (Maclaurin and Sterelny 2008, 114; Bouchard 2013, 92; Nunes-Neto et al. 2014, 124). In ecology, one problem is that functions are regularly attributed to entire groups of organisms or even communities (e.g., Meyer 1993), but it is not clear that such groupings have the right sort of selection history. Another problem is that functions are regularly attributed to abiotic resources, such as soil types, but the selected effects theory has traditionally been restricted to living entities.

Finally, Craver (2013, 154) suggests that he adopts a sort of function pluralism, where different concepts of function may be appropriate in different contexts. But he also seems to assert that selected effects functions are not relevant in neuroscience. As he puts it, "neuroscience and physiology have goals that would be hampered by the general acceptance of such a proprietary notion of function [that is, the selected effects theory]." His argument is that neuroscience requires a notion of function that is neither tied to selection history nor to current adaptiveness: "one can describe the functions of items in the mechanisms for anoxic cell death, the production of cancer, and the progression of Alzheimer's disease" (ibid.; also see Craver 2001, 67, for similar remarks).

As evident from the citations above, various sorts of arguments have been given for accepting between-discipline pluralism. One argument is a sociolinguistic one. This argument claims that the relevant practitioners (e.g., functional morphology, developmental biology, or neuroscience) simply do not use the term "function" to refer to selection history of a trait. Incidentally, we must handle this argument with caution. As noted above (Sects. 3.2 and 4.3) it is very difficult to infer which concept or concepts a practitioner is appealing to by examining overt linguistic usage. A second argument is more metaphysical. For example, Sterelny and Maclaurin (2008, 114) note that ecologists seem to attribute functions to entities that simply do not have selection histories. In other words, ecological communities are just not the right kinds of things to have selected effects functions.

Both sorts of arguments, however, rely on a shared assumption, namely, that the selected effects theory only attributes functions to entities that have undergone natural selection operating over an evolutionary time scale. But, as I noted in Sect. 3.4, many theorists have argued that the sort of selection relevant for attributing functions is much broader than natural selection. It can include, for example, learning by trial and error, antibody selection, and, in my view, neural selection. At the most general level, I believe, the sort of selection that is most relevant for the selected effects theory takes place whenever a trait is differentially replicated, or retained, in a population by virtue of its effects. The selected effects view is applicable whenever function is used in a why-is-it-there sense, in whichever discipline such usage occurs and whether or not natural selection is implicated. Two examples will serve to illustrate the point.

Suppose one asks, "Why does *that* rat press the lever?" An appropriate and perhaps correct answer is, "the function of that rat's lever-pushing is to get food." This is a correct example of a selected effect function but it is correct, I maintain, by virtue of the rat's learning history, rather than by virtue of natural selection. Though natural selection may provide some ultimate explanation of why the rat is capable of learning, that is not, I hold, what grounds the function statement at issue. As I noted in Sect. 3.4, the philosopher of science Israel Scheffler (1959, 269–270) argued that what makes a behavior purposeful is not its current goal directedness, but its history of reinforcement. This is not to say that all learning follows the behaviorist model of trial-and-error (Kingsbury 2008). Rather, I am making the conceptual point that trial-and-error is a process that gives rise to new functions when it occurs, and hence that selected effects functions are not restricted to evolutionary biology.

Another example comes from developmental neuroscience and, in particular, the formation of ocular dominance columns in mammals. In cats, for example, most cells of the visual cortex are responsive to visual information from both eyes. They are called "binocularly-driven." Some cells are responsive only to information coming in from a single eye. These are called "monocularly-driven." The "ocular dominance profile" of a cell refers to the degree to which it is driven by both eyes or by one eye only. The ocular dominance profile of a cell is not entirely independent of its neighbors. Rather, cells with the same ocular dominance profile tend to clump together in bands called "ocular dominance columns." These ocular dominance columns can be visualized using various staining methods. They can be made to appear as a pattern of alternating stripes along the surface of the visual cortex, much like a zebra's stripes.

Here is an interesting result. Suppose one sutures a kitten's eyelid at birth. This procedure is called "monocular occlusion." (I am not going to comment here about the ethical ramifications of this procedure.) Then, the kitten's visual system will develop abnormally. Specifically, within weeks, most of the cells in its visual cortex will be monocularly-driven, and they will be responsive only to light coming in from the unaffected eye. The fact that most cells are monocularly-driven is good for the cat because it maximizes visual acuity in the unaffected eye. Even if one re-opens the sutured eyelid several months later, those cells will no longer be responsive to information coming from the affected eye. That is because the neural connections between the affected eye and the visual cortex have been lost. What happens to them?

A natural thought here is that those neural connections atrophy as a consequence of disuse. But that explanation is not, in fact, correct. After all, if one sutures both eyes of a kitten, one does not see that sort of disuse-related atrophy. Rather, each cell of the visual cortex is the site of a kind of competition between neural connections from the affected eye and the unaffected eye (Wiesel and Hubel 1963, 1015). The repeated activation of the one set of synapses causes the other set of synapses to be lost. This competitive process has been extensively documented and is a paradigm example of synapse selection (see Garson 2011, 2012 for discussion and references).

The fact that the formation of ocular dominance columns is a result of a competitive process is what allows us to assign new functions to neurons in the visual cortex. Why are most cells in *this* kitten's visual cortex monocularly-driven (following monocular deprivation)? Because that arrangement allows those cells to transmit visual information to the rest of the brain. In other words, the reason certain synapses were retained (those connected to the open eye) and others were eliminated (those connected to the sutured eye) is because the former conveyed visual information to the rest of the brain and the latter did not. That is why they are there.

Similar kinds of competitive processes may underlie other sorts of neural plasticity. Consider the way that the somatosensory cortex can reorganize itself after bodily damage such as limb amputation (e.g., Pascual-Leone et al. 2005). In those cases, parts of the cortex that used to serve the amputated limb become recruited, as it were, to serve some other capacity (such as the capacity to detect the position of a prosthetic limb). Some researchers argue that this sort of neural plasticity involves a competitive process at the level of synapses. The idea is that preexisting neural connections are differentially retained, and others eliminated, because of their ability to serve the new capacity (see Rauschecker 1995; Miller 1996; Ramiro-Cortés et al. 2014). If so, then it would be appropriate, in my view, to say that a *function* of the new configuration of synapses is to serve the prosthetic limb. That capacity explains why that particular configuration of synapses exists, that is, why it was retained over some other configuration. It is an appropriate answer to a why-is-it-there question.

If we accept the generalized form of the selected effects theory, how does that affect our understanding of pluralism? First, from an ontological perspective, it means that new functions are emerging on an ongoing basis over the lifetime of the individual and not merely over an evolutionary time frame. Methodologically, it means that scientists may appropriately appeal to selected effects functions in several different disciplines throughout the life sciences and even psychology, and not merely in evolutionary biology. Consequently, I would reject between-discipline pluralism in favor of what one might call "within-discipline pluralism." Within-discipline pluralism is a form of pluralism that emphasizes how different concepts of function can coexist within the same discipline.

References

Aguzzi, A., Sigurdson, C., & Heikenwaelder, M. (2008). Molecular mechanisms of prion pathogenesis. *Annual Review of Pathology, 3*, 11–40.

Allen, C., & Bekoff, M. (1995). Biological function, adaptation, and natural design. *Philosophy of Science, 62*, 609–622.

Amundson, R., & Lauder, G. V. (1994). Function without purpose: The uses of causal role function in evolutionary biology. *Biology and Philosophy, 9*, 443–469.

Bechtel, W., & Richardson, R. C. (1993). *Discovering complexity: Decomposition and localization as strategies in scientific research*. Princeton, NJ: Princeton University Press.

Bock, W. J., & von Wahlert, G. (1965). Adaptation and the form-function complex. *Evolution, 19*, 269–299.

Bouchard, F. (2013). How ecosystem evolution strengthens the case for function pluralism. In P. Huneman (Ed.), *Function: Selection and mechanisms* (pp. 83–95). Dordrecht: Springer.

Brandon, R. N. (2013). A general case for function pluralism. In P. Huneman (Ed.), *Function: Selection and mechanisms* (pp. 97–104). Dordrecht: Springer.

Craver, C. (2001). Role functions, mechanisms, and hierarchy. *Philosophy of Science, 68*, 53–74.

Craver, C. (2013). Functions and mechanisms: A perspectivalist view. In P. Huneman (Ed.), *Function: Selection and mechanisms* (pp. 133–158). Dordrecht: Springer.

Craver, C. F., & Darden, L. (2013). *In Search of mechanisms: Discoveries across the life sciences*. Chicago: University of Chicago Press.

Cummins, R. (1975). Functional analysis. *Journal of Philosophy, 72*, 741–765.

Cummins, R. (1983). *The nature of psychological explanation*. Cambridge, MA: MIT Press.

Darden, L. (2006). *Reasoning in biological discoveries*. Cambridge: Cambridge University Press.

Davies, P. S. (2001). *Norms of nature: Naturalism and the nature of functions*. Cambridge, MA: MIT Press.

Garson, J. (2011). Selected effects functions and causal role functions in the brain: The case for an etiological approach to neuroscience. *Biology and Philosophy, 26*, 547–565.

Garson, J. (2012). Function, selection, and construction in the brain. *Synthese, 189*, 451–481.

Garson, J. (2013). The functional sense of mechanism. *Philosophy of Science, 80*, 317–333.

Garson, J. (Submitted for publication). How to be a function pluralist. *British Journal for the Philosophy of Science.*

Glennan, S. (1996). Mechanisms and the nature of causation.*Erkenntnis, 44*, 49–71.

Godfrey-Smith, P. (1993). Functions: Consensus without unity. *Pacific Philosophical Quarterly, 74*, 196–208.

Griffiths, P. E. (2006). Function, homology, and character individuation. *Philosophy of Science, 73*, 1–25.

Hardcastle, V. G. (1999). Understanding functions: A pragmatic approach. In V. G. Hardcastle (Ed.), *Where biology meets psychology: Philosophical essays* (pp. 27–43). Cambridge, MA: MIT Press.

Hardcastle, V. G. (2002). On the normativity of functions. In A. Ariew, R. Cummins, & M. Perlman (Eds.), *Functions: New essays in the philosophy of psychology and biology* (pp. 144–156). Oxford: Oxford University Press.

Hempel, C. G. (1965). The logic of functional analysis. In C. G. Hempel (Ed.), *Aspects of scientific explanation* (pp. 297–330). New York: Free Press.

Kingsbury, J. (2008). Learning and selection. *Biology and Philosophy, 23*, 493–507.

Kitcher, P. (1993). Function and design. *Midwest Studies in Philosophy, 18*, 379–397.

Machamer, P., Darden, L., & Craver, C. F. (2000). Thinking about mechanisms. *Philosophy of Science, 67*, 1–25.

Maclaurin, J., & Sterelny, K. (2008). *What is biodiversity?*. Chicago: University of Chicago Press.

Matthen, M. (1988). Biological functions and perceptual content. *Journal of Philosophy, 85*, 5–27.

Meyer, C. (1993). Functional groups of organisms. In E. Schulze & H. A. Mooney (Eds.), *Biodiversity and ecosystem function* (pp. 67–96). Berlin: Springer.

Miller, K. D. (1996). Synaptic economics: Competition and cooperation in synaptic plasticity. *Neuron, 17*, 371–374.

Millikan, R. G. (1989). In defense of proper functions. *Philosophy of Science, 56*, 288–302.

Millikan, R. G. (2002). Biofunctions: Two paradigms. In A. Ariew, R. Cummins, & M. Perlman (Eds.), *Functions: New essays in the philosophy of psychology and biology* (pp. 113–143). Oxford: Oxford University Press.

Mitchell, S. D. (1993). Dispositions or etiologies? A comments on Bigelow and Pargetter. *Journal of Philosophy, 90*, 249–259.

Neander, K. (1991). Functions as selected effects: The conceptual analyst's defense. *Philosophy of Science, 58*, 168–184.

Neander, K., Submitted for publication. *The emergence of content: Naturalizing the representational power of the mind.* Cambridge, MA: MIT Press.

Nunes-Neto, N., Moreno, A., & El-Hani, C. N. (2014). Functions in ecology: An organizational approach. *Biology and Philosophy, 29*, 123–141.

Pascual-Leone, A., et al. (2005). The plastic human brain cortex. *Annual Review of Neuroscience, 28*, 377–401.

Piccinini, G., & Craver, C. (2011). Integrating psychology and neuroscience: Functional analyses as mechanism sketches. *Synthese, 183*, 283–311.

Preston, B. (1998). Why is a wing like a spoon? A pluralist theory of function. *Journal of Philosophy, 95*, 215–254.

Prior, E. (1985). What's wrong with etiological accounts of biological function? *Pacific Philosophical Quarterly, 66*, 310–328.

Ramiro-Cortés, Y., Hobniss, A. F., & Israely, I. (2014). Synaptic competition in structural plasticity and cognitive function. *Philosophical Transactions of the Royal Society B, 369*, 20130157.

Rauschecker, J. P. (1995). Compensatory plasticity and sensory substitution in the cerebral cortex. *Trends in Neurosciences, 18*, 36–43.

Sarkar, S. (1998). *Genetics and reductionism*. Cambridge: Cambridge University Press.

Scheffler, I. (1959). Thoughts on teleology. *British Journal for the Philosophy of Science, 9*, 265–284.

Šustar, P. (2007). Neo-functional analysis: Phylogenetical restrictions on causal role functions. *Philosophy of Science, 74*, 601–615.

Šustar, P., & Brzović, Z. (2014). The function debate: Between "cheap tricks" and evolutionary neutrality. *Synthese 191*, 2653–2671. [Check citation??].

Weber, M. (2005). *Philosophy of experimental biology*. Cambridge: Cambridge University Press.

Wiesel, T. N., & Hubel, D. H. (1963). Single-cell responses in striate cortex of kittens deprived of vision in one eye. *Journal of Neurophysiology, 26*, 1003–1017.

Wimsatt, W. C. (1976). Reductive explanation: A functional account. In R. S. Cohen & A. Michalos (Eds.), *Proceedings of the 1974 meeting of the Philosophy of Science Association* (pp. 671–710). Dordrecht: D. Reidel.

Wimsatt, W. C. (1986). Forms of aggregativity. In A. Donagan, A. N. Perovich, & M. V. Wedin (Eds.), *Human nature and natural knowledge* (pp. 259–291). Dordrecht: Reidel.

Chapter 6
Alternative Accounts of Function

Abstract In this chapter, I consider three theories of function that are relatively new, in the sense that they have been developed over the last twenty years. The "weak etiological theory" says, roughly, that a trait token in an organism has a function so long as that kind of trait contributed to the fitness of that organism's ancestors and it is inherited. It defines function in terms of inheritance and past contributions to fitness, but not selection. I assess some differences between this theory and the standard selected effects account and question the motivation for the account. A second group of theories is known as the "systems-theoretic" or "organizational" view. This is not a single theory but a family of theories based on the idea that a trait token can acquire a function by virtue of the way that very token contributes to a complex, organized, system, and thereby to its own continued persistence, as a token. I argue that the organizational approach faces liberality problems. Finally, the modal theory of function holds that the function of a trait token has to do with the behavior of that token in certain nearby possible worlds. I assess the theory and survey some problems. Bence Nanay developed the modal theory as an attempt to solve a certain circularity problem that he believes afflicts most other theories of function, but it is not clear whether there is a real problem here to be resolved.

Keywords Weak etiological theory of function · Organizational functions · Systems-theoretic functions · Modal theory of function · Biological trait

6.1   The Weak Etiological Theory

This chapter will discuss three alternative theories of function that arose in the last twenty years, or that have gained fresh support over that time. The first two can be seen, in some respects, as contenders to the selected effects theory. More precisely, the first two agree that function ascriptions, at least sometimes, purport to be causal explanations for a trait's existence. They agree that when biologists attribute functions to items, they are attempting to account for why they are there, in the

© The Author(s) 2016   97
J. Garson, *A Critical Overview of Biological Functions*,
Philosophy of Science, DOI 10.1007/978-3-319-32020-5_6

causal sense of explanation. However, they do not believe that reference to selection, per se, is necessary for functions to play this role.

The first of these is the "weak" etiological theory. Buller (1998) presented this as an alternative to the selected effects theory (which he called the "strong" etiological theory). The weak etiological theory says, roughly, that a trait token in an organism has a function so long as that kind of trait contributed to the fitness of that organism's ancestors and it is inherited. For example, it says that Amadi the zebra's stripes have a function because Amadi's parents had stripes, stripes benefited his parents, and they are inherited. It defines function in terms of inheritance and past contributions to fitness. It does not define function in terms of selection.

How does one appeal to past contributions to fitness without also appealing to selection? I will clarify his definition to show how this can be done. The weak theory holds that a type of trait T has function F in organism O under the following conditions:

1. some amongst O's ancestors had T;
2. T did F in those ancestors;
3. T's doing F in those ancestors contributed to the fitness of those ancestors; and
4. T is inherited.

Now, importantly, Buller construes fitness as a non-comparative notion, as I explained in Sect. 4.2. To say that a trait "contributes to fitness" in this sense is to say that it makes a causal contribution to some component of fitness, namely, "viability, fertility, fecundity, or mating ability" (roughly, survival or reproduction). He does not think that the notion of a "contribution to fitness" is a relative notion, that is, that it makes implicit reference to *other* alternative traits. My lungs help me breathe, and breathing helps me survive. This would be true even if there were no variation among human populations for breathing ability and breathing ability is not undergoing selection. In theory, a trait may have gone to fixation by random genetic drift, but later, only after going to fixation, it may have started helping its bearers survive. For example, it is at least logically possible that all zebras came to have stripes as a result of random genetic drift, and much later, their habitats became infested with Tsetse flies, and the stripes started helping them survive and reproduce, but stripes were never selected for. The traditional selected effects theory would have to deny functions to such traits. Buller's weak etiological theory could assign functions to those traits.

This view is supposed to be explanatory, in the straightforward sense of giving a causal explanation for the existence of T in O. Suppose we want to know why Amadi the zebra has stripes. In Buller's view, part of the explanation has to do with what the stripes did in Amadi's parents (or earlier ancestors). Stripes helped Amadi's parents ward off biting flies. That helped them survive until maturity and successfully breed. Amadi is amongst their brood. Because of the mechanisms of inheritance, Amadi, like his parents, has stripes. So, part of the explanation for why Amadi has stripes appeals to something that the stripes did in the past (in his parents). Function ascriptions, therefore, are causal explanations. We can make sense of the explanatory dimension of functions without invoking selection.

Buller takes this argument a step further. He thinks that even in the context of the selected effects theory of function, selection per se *is not actually doing the explanatory work* (520). If he is right, that would be a major embarrassment to the selected effects theory, because it would suggest that selected effects theorists have made a significant error regarding which part of their view is doing the explaining. The fact that there was selection for stripes does not explain why Amadi has stripes. Rather, what explains why Amadi has stripes is the fact that stripes helped his ancestors breed and they are inherited. In a sense, this goes back to the point that Cummins (1975) made against the selected effects theory. If we are trying to explain why an individual has a property, the fact that there was selection for that property is causally irrelevant. As Sober (1984) put the point, selection does not play a role in "developmental" explanations. (As I noted there, this claim is controversial but I will let it stand.)

On the face of it, Buller's view seems very attractive. It appears to capture the explanatory dimension of function statements, but it does so by appealing to a broader range of biological processes than selection alone. But I have three concerns with Buller's view. First, Buller's claim that selection is not doing any explanatory work relies on a very specific, and I think questionable, assumption about what the *explanandum* of the function statement is. To my ears, when we state, "the function of zebra stripes is to deter flies," we are trying to explain why zebras have stripes. In other words, we are first and foremost asking a population-level question. We are not asking, first and foremost, why Amadi has stripes. In order to answer the *population-level* question, it is necessary to appeal to selection (assuming that selection was, in fact, involved). In other words, the reason that zebras have stripes (as opposed to, say, being monocolored like horses) is because there was selection for stripes in a population of ancestors to modern zebras. Buller (1998, 300) grants that *if* the *explanandum* of the function statement is some population-level fact, *then* selection might be part of the explanation for that fact, but he denies that function statements purport to explain population-level facts.

Buller and I could quibble about what, precisely, functions are supposed to explain. Even if we set that aside, I have two other concerns. The second has to do with the motivation for the view. I am not entirely clear about what the motivation for the weak etiological view is. One of the concerns that Buller has is that, if our theory of function appeals to selection, then that theory will be overly limited. A lot of traits will turn out not to have functions when they intuitively do. As he puts it, the selected effects theory lets us assign functions in the context of evolutionary biology and behavioral ecology. But, in many cases, it does not let us assign functions in the context of anatomy and developmental biology. He thinks that, in those contexts, there are a host of traits that make contributions to fitness, but that were never selected for (523).

But why does he think this? He says that a lot of traits that developmental biologists are interested in do not vary much. Without variation, there is no selection. He claims that studies of natural selection in the wild have yielded the following result: there is "a far greater incidence of variation in morphological traits than in physiological and biochemical traits…if there is a very low incidence of variation in physiological and biochemical traits, there is perhaps an even lower incidence of variation among such a traits within common selective environments" (512–513). The selected effects theory cannot assign functions to these traits.

I do not think Buller offers strong evidence for the claim that there are many interesting developmental traits that contribute to fitness, but have never undergone selection. Here is one consideration. Just because we see little variation for a trait, that does not mean there is no selection for it. Rather, it could be undergoing intense stabilizing selection. Consider neurulation, the process responsible for the development of the spinal cord. We see almost no variation for this process across mammals. That is because disorders of neurulation tend to be fatal. So current examination of those traits will reveal little variation. But that does not mean that neurulation was not selected for! If anything, the exact opposite is the case. It was selected for and it is undergoing strong stabilizing selection.

A third and final concern stems from the fact that Buller assumes, without argument, that functions can only be attributed, in the first place, to entities that are capable of reproducing. Traditional versions of the selected effects theory, for example, Millikan's and Neander's, also made this assumption. But as I indicated above (Sect. 3.4), I believe this assumption is somewhat unprincipled and arbitrary. Why should functions be so restricted? Buller claims that his view is superior to the selected effects theory because it does away with arbitrary restrictions (namely, the appeal to selection), but his view arguably contains arbitrary restrictions.

6.2   Systems-Theoretic Functions

A second group of theories is known as the "systems-theoretic" or "organizational" view. The basic idea is this. Go back to Buller's view for a moment. He points out that some traits contribute to their own intergenerational reproduction by contributing to the fitness of the organism that has them. For example, the zebra's stripes contribute to their own intergenerational reproduction by deterring flies, and thus helping the individual live long enough to reproduce. Since they are inherited, when they confer such a fitness advantage they indirectly ensure their own perpetuation through the generations. Theoretically, stripes can contribute to their intergenerational reproduction even if there is no selection taking place.

But some trait tokens contribute not (or not merely) to their intergenerational reproduction, but to *their own persistence, as tokens, within the individual*. They do so by contributing to the functioning of the organism as a whole. Organizational (or systems-theoretic) analyses of function take this insight as their starting point, though they flesh it out in different ways. In short, the idea is that a trait token can come to have a function simply by how it contributes to its own persistence in the organism. (This may sound a lot like my own view, which emphasizes the differential persistence of a trait—see Sect. 3.4. But there is a crucial difference: my view insists that *selection processes*, broadly construed, are necessary for functions, but the systems-theoretic view does not.)

Consider the heart. The heart beats; in doing this, it circulates blood; in doing this, it brings nutrients to the cells of the body, including those that constitute the heart itself. So, in beating, it contributes to its *own* persistence in the organism. This observation suggests a disjunctive account of function. Cast in a forward-looking vein, we can say that trait T possesses function F either when T's doing F contributes to the intergenerational reproduction of T (as in the zebra's stripes), *or* when T's doing F contributes to the intra-generational persistence of T (as in the heart's beating). (Clearly, the "or" should be understood inclusively, since the heart's pumping contributes not only to its intergenerational reproduction but also its intra-generational persistence.) Schlosser (1998) is the first paper to work out this core idea in a systematic manner, and similar views were advocated by McLaughlin (2001), Christensen and Bickhard (2002), Weber (2005), Mossio et al. (2009), Saborido et al. (2011), and Moreno and Mossio (2015). Sarkar (2005, 18) develops a view of function in which the function of a trait has to do with its contribution to the persistence of the system as a whole. I think this is slightly different from the systems-theoretic view, because Sarkar does not demand that the component in question, by contributing to the persistence of the system, must also contribute to *its own* persistence as a component.

Conceptually, this systems-theoretic view of function comes in at least two forms, backwards-looking and forward-looking. First, one might hold that the function of a trait results from what it did *in the past* that contributed to its persistence or reproduction (the etiological version). Second, one might hold that the function of a trait has to do with its current-day disposition to contribute to its *future* persistence or reproduction (the forward-looking version). Schlosser (1998) prefers the forward-looking approach, as do, I believe, Christensen and Bickhard (2002). McLaughlin (2001) clearly prefers the etiological formulation.

Mossio et al. (2009), Saborido et al. (2011), and Moreno and Mossio (2015) suggest a neutral, "atemporal" approach—that is, neither backwards nor forward looking. The idea is that a part of a system has a function because of the way the system, as a whole, instantiates a certain abstract pattern of functional dependencies (Mossio, pers. comm.). I think it would be wise for them to consider taking a more backwards-looking perspective, however, since the backwards-looking perspective helps us account for how functions are explanatory, which is what they want. I have a hard time seeing why the fact that a system realizes an abstract system of functional dependencies actually *explains* why the system or any of its parts exist. A reason my heart exists now is because of something it did in the past and not because it exemplifies a certain pattern of relationships.

Suppose we accept an etiological reading of the systems-theoretic view, where a trait token can acquire a function by virtue of its past contribution to its own persistence. Such an account would seem to account for how functions can be explanatory, and it would do so without appealing to selection history (and without even appealing to how it contributed to the fitness of ancestors). That would make it a strong contender to the selected effects theory of function. Consider the following: why do I have a heart, right now, at this instant? Well, part of the explanation is that my heart circulated blood yesterday, and its circulating blood yesterday helped to bring nutrients to the cells of my body, and hence contributed to rebuilding and repairing the cells of my heart itself. So, the fact that my heart was beating yesterday partly explains why I have a heart now. If this is correct, then the systems-theoretic view captures the explanatory dimension of function statements in a way that is even more liberal than both the traditional selected effects view or the weak etiological theory (e.g., Moreno and Mossio 2015, 70).

I think this theory faces a significant problem, namely, the theory is threatened by the same sort of liberality objection that faced Wright's (1973) initial articulation of the etiological theory of functions (as detailed in Sect. 3.2). According to Wright, the function of a trait is just some effect that explains why it is there. In principle, Wright's extremely general formulation can apply to trait tokens (as in the systems-theoretic view) and trait types. Yet Boorse (1976) showed that Wright's formulation led to a series of famous counterexamples. Obesity contributes to a sedentary lifestyle, which, in turn, promotes obesity. So obesity explains, to this extent, why it is there, that is, why a given individual has remained obese. But that is not its function. If, on the systems-theoretic view, a trait can acquire a new function simply by contributing to its own persistence in an individual, then the counterexamples remain.

Schlosser (1998, 312) tries to avoid the Boorse-type counterexamples by imposing a complexity constraint. He states that, in order for a trait to have a function, the trait must utilize *diverse* mechanisms or means for achieving its own persistence or reproduction. Consider Boorse's leaking hose that knocks out scientists. The hose springs a leak (*T*) which knocks out scientists (*F*) and thereby contributes to the persistence of the leak (*T*). But, according to Schlosser, the leak does not have a function, because that particular cause-and-effect sequence is not complex enough, or it is not complex in the right way. In contrast, consider the extremely diverse behavioral mechanisms that the mimic octopus (*Thaumoctopus mimicus*) uses to bring about the function of camouflage (Norman et al. 2001). It can mimic several different kinds of living creatures, including flatfish, lionfish, and sea snakes, depending on the specific circumstances at hand.

It does not seem to me that Schlosser's complexity criterion achieves its stated aim. A very interesting example of a complex feedback process can be found in the recurrence of panic attacks. Cognitive behavioral therapists have shown that there are several ways that the experience of a single panic attack can set the stage for the recurrence of such attacks (Clark 1997). One way is by promoting, in the person, a heightened vigilance to bodily sensations. Another way is by inducing the person to avoid precisely those situations that would have the effect of *disconfirming* the person's mistaken beliefs. (For example, if a person avoids jogging out of the belief that it can cause a heart attack, then the person is deprived of opportunities to disconfirm the mistaken belief.) So a panic attack is a component of a "complex self-reproducing system." That is, panic attacks are self-reproducing and they use diverse "means" to achieve this effect. But I take it that panic attacks do not have the function of promoting a heightened vigilance to bodily sensations. Similar remarks can be made about the persistence of obesity. There are many ways that obesity contributes to its own persistence. It makes physical exertion more exhausting. It makes some people feel embarrassed to go to a gym and work out.

Schlosser (pers. comm.) does not think that the example of obesity is a serious problem. In his view, the appropriateness of function ascriptions is a matter of degree, and is proportional to the number of different pathways by which a trait can contribute to its own reproduction. Hence, the fact that the relation between obesity and a sedentary lifestyle can be mediated by *two* series of state transitions rather than merely *one* makes it only slightly less trivial to ascribe a function to obesity than it would be to ascribe a function to junk DNA, which can only reproduce itself by a single pathway. Moreover, since living systems, with their exuberant complexity, have evolved from simpler, non-living systems, one should not expect a sharp division between appropriate and inappropriate function ascriptions.

Clearly, the adequacy of Schlosser's response hinges on the assumption that there are no more than two, or at least a very few, circumstances in which (say) obesity leads to a sedentary lifestyle or panic to mistaken beliefs. However, the relation between obesity and a sedentary lifestyle can be as complex as the techniques of avoidance that the human mind is capable of contriving. Under one circumstance (e.g., embarrassment about physical appearance), obesity is necessary for producing aversion to, say, purchasing a membership at a gym; under another (e.g., fear of excessive perspiration), obesity is necessary for producing aversion to, e.g., staying outdoors for long periods of time; under a third (e.g., production of fatigue), obesity is necessary to bringing about the early cessation of strenuous physical activity, and so on. Clearly, one could continue to generate such scenarios, and hence an arbitrarily large number of independent trajectories can mediate the relation between the two, making the relation between obesity and a sedentary lifestyle complex to an arbitrarily large degree. The same can be said of the relation between panic and mistaken beliefs.

In a sense, the problem is one of grain. When we describe a system as "complex," we assume some way of dividing up a system into parts or components and numbering them (McShea and Venit 2001; McShea and Brandon 2010). Perhaps there is some sense in which, in my obesity example, I'm not getting to the right sort of complexity, or the sort that matters to biology. Maybe if we had a firmer grasp on the notion of complexity we would see that the examples of panic attacks or obesity do not exhibit genuine complexity. We can hope that future research will yield a more definitive judgment.

Mossio et al. (2009) do not explicitly discuss the Boorse-type counterexamples, but they make some remarks that would seem to avoid them. They first describe a class of systems they call "self-maintaining" systems. These self-maintaining systems are ubiquitous throughout the physical world. These are systems that promote their own continued existence. They refer to this property, following Varela (1979) as "closure." Candle flames and hurricanes are examples of such systems. But we do not want to ascribe functions to hurricanes. That would be too liberal. So, what else is needed for functions?

They recommend that we attribute functions only to a subset of self-maintaining systems, namely, those that exhibit "organizational differentiation." These are systems in which we can "distinguish between different contributions to self-maintenance made by the constitutive organization," and each component makes a "specific contribution to the conditions of existence of the whole organization" (826). An organizationally differentiated system is one with different kinds of parts, and in which these different parts contribute differently to self-maintenance. The human body is a great example of an organizationally differentiated system. (Note that these definitions have grown more complex and sophisticated since Mossio et al. (2009). Chapter 3 of Moreno and Mossio (2015) is the best recent introduction to their approach for those who wish to follow out this trail of thinking.)

I do not think that this maneuver will avoid all of the Boorse-type counterexamples (nor was it designed to). For it seems to me that the phenomenon of panic attacks illustrates organizational differentiation. Imagine a young man, Naeem, who is susceptible to panic attacks. Inside Naeem there is a variety of psychological dispositions that work together to promote panic attacks. For example, one of these dispositions is that Naeem is hypervigilant to bodily sensations. Another disposition is that he is averse to strenuous exercise (in the fear that it will cause a heart attack), and hence he avoids the situations that would disconfirm his fears. These are two components of a psychological "system," which we can call the "panic attack system." It seems to me that this system exhibits "closure," in that it promotes its own existence, and that it exhibits "organizational differentiation," since there are different parts that do different things that help the system as a whole to persist. But it does not seem to me that hypervigilance to bodily sensations has the function of producing panic attacks. It is not clear to me how one would avoid this implication without introducing a host of potentially ad hoc modifications. This is particularly worrisome as Saborido and Moreno (2015) wish to apply the organizational view to biomedicine and pathology.

One way the organizational theorists may try to avoid this implication is to restrict functions to the components of *organisms*, and to assign functions on the basis of the way the item contributes to the functioning of the organism as a whole. This would make sure that the parts of the panic system do not have functions. But I think this restriction to organisms would go against the whole point of the systems-theoretic view. As I understand it, the point of the view is to give us a very abstract account of system-hood that would apply to several different sorts of systems, such as ecosystems and machines, and not just organisms.

McLaughlin (2001) also advances a systems-theoretic view. He attempts to avoid the Boorse-type counterexamples by introducing a value component. His view, then, is a type of evaluative approach to function (which I described briefly in Sect. 3.3). He claims that the reason clay crystals do not have functions is because there is no beneficiary (168). As I noted there, however, this evaluative approach raises fairly deep metaphysical and meta-ethical questions that we may wish to avoid. If functions are evaluative, are values subjective or objective? If values are subjective, then functions are not, contrary to appearance, mind-independent facts about reality. If values are objective, then how do they fit into the natural world?

6.3 The Modal Theory of Functions

Finally, I turn to the modal theory of function. The modal theory of function is motived by a puzzle about how biologists individuate trait types. My heart, your heart, and President Obama's heart are all tokens of the same biological category, *the heart*. But by virtue of *what* do these individual tokens belong in the same category? Is it because of their shared morphology? Is it because of their shared causal role? Is it because they are homologous (that is, descended from the same ancestral heart token)? Nanay (2010) joins other philosophers of biology, such as Rosenberg and Neander (2009), in the view that all of these answers are wrong. What makes a token heart a member of the biological category *heart* is that it has the function of circulating blood. We can summarize this thesis in a slogan: "trait type individuation is by function." But if that is right, then when we are constructing a definition of function, we are forbidden, on pain of circularity, from appealing to the notion of a trait type. Hence, Nanay concludes, most theories of function are mistaken.

A solution to this puzzle is to define "function" so that it applies first and foremost to trait *tokens*: my heart, Amadi's stripes, and so on. The function of Amadi's stripes must be based on properties of that very token. Nanay thinks functions should be defined in terms of certain *counterfactual* states of affairs involving that token. Assuming that the best way of thinking about counterfactual situations is in terms of possible worlds, the modal theory holds that the function of a trait token has to do with the behavior of that token in certain nearby possible worlds.

What must we do in order to determine whether a trait token in an individual has a function (e.g., trait t in organism o has function f)? We first consider all of the possible worlds in which o has t and t does f. (All of these worlds must be "relatively close," but within these relatively close worlds, some are closer than others. A world where eagles breathe fire onto approaching predators is not a close world for our purposes.) If, in the nearest possible world in which o has t and t does f, t's doing f contributes to o's fitness (and if that world is closer to ours than any world in which t's doing f *doesn't* contribute to o's fitness), then the function of t (in our world) is f.

An example should help to clarify the point. Suppose Sam the eagle has a broken wing. Does Sam's wing have the function of flight? We look at those nearby worlds in which Sam's wings help him fly, and where flying contributes to fitness. We also look at those nearby worlds in which Sam's wings help him fly but that does not contribute to fitness. If, on the nearest possible world where Sam's wing enables flight, his flying contributes to fitness, *and* that world is closer to ours than any world in which Sam's wing enables flight but flying does not contribute to fitness, then we can say that Sam's wing (in the actual world) has the function of flight.

There are three problems with this view. The first is that it is not clear how to measure the closeness of various possible worlds. In other words, there is an inherent indeterminacy about evaluating the relevant counterfactuals. (This problem came up in Sect. 4.2 when we considered what it is for a trait to contribute to fitness.) The second problem is, even if we accept the legitimacy of this idea, the theory generates counterintuitive function ascriptions (Neander and Rosenberg 2012; Artiga 2014). For example, my nose holds up my glasses and this contributes to my fitness (by helping me not get hit by cars). So, on the nearest possible world where my nose holds up glasses, its doing so contributes to fitness. So that is its function. Nanay's (2012) response to both of these problems is the same. He says that what counts as a close world depends on our explanatory purposes. So, on his view, like the causal role view, whether a trait has a function or not, and which function it has, depends on our explanatory goals. Nanay seems to avoid the problem by invoking a pragmatic, "perspectival" approach to function.

Third, and most importantly, the alleged puzzle that is supposed to motivate this theory of function does not strike me as compelling. But unless it is compelling, it is hard to see why we should accept the whole package. First, the claim that *all* trait types are individuated by function strikes me as implausible on its face. That is because it would seem to imply that biologists have no way to classify non-functional trait types, such as freckles, chin clefts, or pattern baldness. But they do. Such considerations have led many philosophers of biology to reject the idea that traits are generally typed by function (Amundson and Lauder 1994; Griffiths 2006). Nanay has a ready response here. Even if we do not accept that *all* traits are classified by function, at least *some* traits are, such as hearts and wings. So one could make the more modest claim that, so long as some traits are classified by function, if we wish to attribute functions to them, we must accept something like the modal theory.

Neander and Rosenberg (2012) make the interesting response that, though there is an intimate relationship between trait-type identification and functions, there is no deep circularity problem. *If* we define "trait-type" in terms of function, and "function" in terms of trait-type, then there would be a problem. Instead, they argue, both "function" and "trait type" are defined in terms of the underlying notion of a *lineage of trait tokens parsed by selection events*. In other words, they are willing to grant that all trait types have functions, but "trait type" is neither defined in terms of "function," nor vice versa.

Nanay (2012) rejects this solution because he says that the very idea of a *lineage of trait tokens* presupposes some way of individuating trait types, and therefore that Neander and Rosenberg's solution is still circular. Another way of putting the point is the say that the very notion of a lineage of trait tokens presupposes an account of what Griffiths (2006, 19) calls "trans-generational character identity," which poses a kind of trait-individuation problem. Nanay thinks that the only way to solve the problem of trans-generational character identity is to appeal to functions.

But I think that Neander and Rosenberg's solution has some merit. It seems to me that one can define the idea of a *lineage of trait token* without developing a particularly rich theory of trait type identification. For example, in order to define the notion of a lineage of trait tokens, we may use Ruth Millikan's (1984, 19–20) notion of "copying." This notion, in turn, is defined in terms of causation and counterfactuals. Roughly, *B* is a copy of *A* if *A* and B have some properties in common, *A* causes *B* to have those properties, and, if *A* had differed in certain respects, *B* would have differed in certain respects as well. It does not require an antecedent account of what constitutes a trait type. Griffiths (2006, 19) also suggests that Millikan's account here is sufficient to give us a theory of trans-generational character identity, and I think this is right. So, here is a procedure for solving the circularity problem: first, use Millikan's notion of a reproductively established family to provide an account of a lineage of trait tokens. Then, use that account, as Neander and Rosenberg (2012) do, to define *both* the notion of a trait-type as well as the notion of function. At any rate, Nanay's view on function provides us with another example of how questions about the nature of function lead us very quickly into other foundational concepts in the philosophy of biology, and in the philosophy of science more generally.

References

Amundson, R., & Lauder, G. V. (1994). Function without purpose: The uses of causal role function in evolutionary biology. *Biology and Philosophy, 9*, 443–469.

Artiga, M. (2014). The modal theory of function is not about functions. *Philosophy of Science, 81*, 580–591.

Boorse, C. (1976). Wright on functions. *Philosophical Review, 85*, 70–86.

Buller, D. J. (1998). Etiological theories of function: A geographical survey. *Biology and Philosophy, 13*, 505–527.

Christensen, W. D., & Bickhard, M. H. (2002). The process dynamics of normative function. *The Monist, 85*, 3–28.

Clark, D. M. (1997). Panic disorder and social phobia. In D. M. Clark & C. G. Fairburn (Eds.), *Science and practice of cognitive behaviour therapy* (pp. 119–153). Oxford: Oxford University Press.

Cummins, R. (1975). Functional analysis. *Journal of Philosophy, 72*, 741–765.

Griffiths, P. E. (2006). Function, homology, and character individuation. *Philosophy of Science, 73*, 1–25.

McLaughlin, P. (2001). *What functions explain: Functional explanation and self-reproducing systems*. Cambridge: Cambridge University Press.

McShea, D. W., & Brandon, R. N. (2010). *Biology's first law*. Chicago: University of Chicago Press.

McShea, D. W., & Venit, E. P. (2001). What is a part? In G. Wagner (Ed.), *The character concept in evolutionary biology* (pp. 259–284). San Diego: Academic Press.

Millikan, R. G. (1984). *Language, thought, and other biological categories*. Cambridge, MA: MIT Press.

Moreno, A., & Mossio, M. (2015). *Biological autonomy: A philosophical and theoretical inquiry*. Dordrecht: Springer.

Mossio, M., Saborido, C., & Moreno, A. (2009). An organizational account for biological functions. *British Journal for the Philosophy of Science, 60*, 813–841.

Nanay, B. (2010). A modal theory of function. *Journal of Philosophy, 107*, 412–431.

Nanay, B. (2012). Function attribution depends on the explanatory context: A reply to Neander and Rosenberg's reply to Nanay. *Journal of Philosophy, 109*, 623–627.

Neander, K., & Rosenberg, A. (2012). Solving the circularity problem for functions. *Journal of Philosophy, 109*, 613–622.

Norman, M. D., et al. (2001). Dynamic mimicry in an indo-malayan octopus. *Proceedings of the Royal Society of London B 268*, 1755–1758.

Rosenberg, A., & Neander, K. (2009). Are homologies (selected effect or causal role) function free? *Philosophy of Science, 76*, 307–334.

Saborido, C., & Moreno, A. (2015). Biological pathology from an organizational perspective. *Theoretical Medicine and Bioethics, 36*, 83–95.

Saborido, C., Mossio, M., & Moreno, A. (2011). Biological organization and cross-generation functions. *British Journal for the Philosophy of Science, 62*, 583–606.

Sarkar, S. (2005). *Molecular models of life*. Cambridge, MA: MIT Press.

Schlosser, G. (1998). Self-re-production and functionality: A systems-theoretical approach to teleological explanation. *Synthese, 116*, 303–354.

Sober, E. (1984). *The nature of selection*. Chicago: University of Chicago Press.

Varela, F. (1979). *Principles of biological autonomy*. Elsevier: North Holland.

Weber, M. (2005). *Philosophy of experimental biology*. Cambridge: Cambridge University Press.

Wright, L. (1973). Functions. *Philosophical Review, 82*, 139–168.

Chapter 7
Conclusion: What Next?

Abstract This chapter begins by showing why one standard way of describing the functions debate—as a debate between the selected effects theory, fitness-contribution theory, and causal role theory—is misguided. I then summarize three main conclusions. First, I argue that, to the extent that function statements are causal explanations, there are no current, viable alternatives to the selected effects theory. Second, I argue that, to the extent that we accept pluralism, we should not accept what I call (in Chap. 5) "between-discipline pluralism," which seeks to restrict the applicability of the selected effects theory to some branches of evolutionary biology. Third, I advocate a specific version of the selected effects theory, the generalized selected effects theory, which I described earlier (in Chap. 3). In closing, I outline three outstanding problems for most theories of function. The first is function indeterminacy. Typically, there are a number of different activities associated with the performance of a function and most theories of function do not have the resources to specify which of those activities, precisely, constitutes an item's function. Second, in many cases, a trait not only has a function, but it has an appropriate rate of functioning. For example, the heart not only has the function of beating, but it has the function of beating at a certain rate. How should we specify the appropriate rate of function? Third, for many traits that have functions, we can distinguish between appropriate and inappropriate situations for the performance of those functions. How should we modify our theories to take this into account?

Keywords Function pluralism · Situation-specificity · Response functions · Rate of functioning · Function indeterminacy

I hope this volume has shown that the functions debate, far from being settled, is alive and well. The terms of the debate, however, have changed substantially over the last twenty years. About twenty years ago, there were three canonical theories: the selected effects theory, the fitness-contribution theory, and the causal role theory. The debate had to do with which of those three theories was superior, or whether we should accept a pluralistic outlook instead. Now, we can see that the framework of that traditional debate is misguided, for two reasons.

© The Author(s) 2016
J. Garson, *A Critical Overview of Biological Functions*,
Philosophy of Science, DOI 10.1007/978-3-319-32020-5_7

First, there are new theories in the mix that we must confront, including the organizational theory, the modal theory, and the generalized selected effects theory. How do those fit into the pluralistic formula? Should these new theories simply be embraced within a wider, pluralistic, outlook? Should they be rejected? Should one of these newcomers replace one of the canonical theories?

Second, there is not just one etiological theory, but at least four different versions of the theory that deserve attention. The traditional or "strong" etiological theory holds (roughly) that the function of a trait is that activity that contributed to the differential reproduction of that trait. The "weak" etiological theory holds that the function of a trait token in an organism is that activity that past instances of that token produced that contributed to the fitness of that organism's ancestors (even if it did not contribute to their differential reproduction). The generalized selected effects theory holds that the function of a trait has to do with its past contribution to either the differential reproduction or the *differential persistence* of the trait within a population. This view is an extension of the selected effects theory because it emphasizes the role of selection (broadly construed) but it rejects the requirement for reproduction. The organizational view (or at least its etiological rendition) holds that the function of a trait is simply its past contribution to its own reproduction or even its own persistence within an individual, independent of selection. I hope that this volume motivates philosophers to pursue these questions in a more rigorous way.

Not only is the functions debate alive and well, but it is just as relevant to the sciences, and to broader philosophical concerns, as it has ever been. Recent debates about the concept of function in genetics, particularly with the advent of the ENCODE project consortium, cry out for more philosophical attention than they have received. In neuroscience, the development and refinement of optogenetics (Emiliani et al. 2015) promises to reveal in greater depth than was previously attainable the functional properties of neurons. Such technologies call for more philosophical reflection about the nature of functions in neuroscience. Ongoing debates in the conservation community about the nature and value of ecological restoration would benefit greatly from philosophical reflection on how functions work in ecology. The functions debate is also relevant to ongoing questions in biomedicine and psychiatry about the right way to classify physical and mental disorders.

Although this volume is primarily intended as an up-to-date survey of the field, there are three main points, or conclusions, that I wish to advance. First, I hope to have shown that, to the extent that functions are explanatory, and to the extent that we understand this demand for explanation in a causal way, there are currently no viable alternatives to the selected effects theory of function (construed broadly to include the generalized selected effects theory). The only alternatives that purport to capture the causal-explanatory sense of function are the weak etiological theory and the organizational theory, and I have outlined what I see as the main problems for those theories in the previous chapter.

Second, I believe that some form of function pluralism is a reasonable and conciliatory stance. However, I do not accept one particular version of that pluralism, which I label "between-discipline" pluralism. Between-discipline pluralism maintains that different theories of function are appropriate for different scientific disciplines. As I have detailed in Sect. 5.3, I believe this view is motivated by an overly narrow construal of what the selected effects theory actually holds. Instead, to the extent that we wish to be pluralists, we should accept what I call "within-discipline" pluralism. Within-discipline pluralism seeks out and acknowledges the plurality of senses of function within any particular scientific discipline.

Finally, I have taken the opportunity here to advance a particular version of the selected effects theory, the generalized selected effects theory. The traditional selected effects theory, which we primarily owe to Neander (1983) and Millikan (1984), imposes an unprincipled restriction on the sorts of entities that can possess (direct proper) functions. According to this restriction, in order for an entity to have a function it must be the sort of thing that can reproduce. But, e.g., synapses do not reproduce. They simply persist better or worse than other synapses. In my view, the function of a trait is that activity that, in the past, caused the differential reproduction *or* the differential persistence of the trait within a biological population. This version does away with an unnecessary restriction on the traditional selected effects theory and it can also handle various liberality objections.

In closing, I will outline some outstanding theoretical problems that any theory of function must confront. I believe that, to some extent, these problems can be tackled in a theory-independent way. What I mean is that these problems affect all of the different approaches to function to some degree or another and it is possible that the solutions to these problems can be articulated in a very general way, that is, in a way that could be applicable to any specific theory of function.

There are at least three problems: function indeterminacy, appropriate rate of function, and situation-specificity of function. The first problem is that of function indeterminacy and, specifically, what can be called the "hierarchical" form of indeterminacy. The problem is simple. Any given trait (say, the heart) performs its function by triggering a series of events. For example, the heart beats. In beating, it circulates blood; in circulating blood, it brings energy to cells and eliminates waste. In doing so, it keeps us alive and healthy. So the question is, *which of these events*, specifically, constitutes its function? Is the determinate function of the heart merely to beat? To circulate blood? To bring energy to cells? Or something else?

Many philosophers of biology tackled the problem of indeterminacy in the 1980s and 1990s, but no particular consensus was achieved (e.g., Millikan 1984, 34; Dretske 1986; Sterelny 1990; Neander 1995; Goode and Griffiths 1995; Buller 1997). It is tempting to dismiss the problem as a philosophers' worry. After all, who cares which of these activities we pick out as constituting the heart's function? But that would be an error. Solving function indeterminacy is relevant for solving other philosophical problems. One area has to do with extending the notion of biological function into the philosophy of mind, to think about representational content (for an overview, see Garson 2015 Chap. 7). Another problem is that, strictly speaking, if we cannot specify what the determinate function of a trait is, we cannot say, precisely, when that trait is dysfunctional or malfunctioning (see Garson Submitted for publication for discussion). After all, a trait may be able to carry out one of the activities in the sequence and not another. Whether or not the trait is dysfunctional depends upon which event in that sequence constitutes its function.

A second problem is this: even if we know what the function of a trait is, there is a further question of what its appropriate *rate* of functioning is. A heart that cannot beat at all, for example, during cardiac arrest, is dysfunctional. But a heart that is beating at an unacceptably low rate, for example, with an ejection fraction of 20 %, is also dysfunctional. So, when we judge that a trait is functional (with respect to a specific activity, such as the heart's beating) we are also making a judgment to the effect that it is beating *at its appropriate rate*. So what constitutes an appropriate rate of function? Though Boorse (2002, 71) recognized the problem, very few theorists have considered this problem in any real depth. Schwartz (2007), Hausman (2012), and Garson and Piccinini (2014) attempt to formulate answers.

There is a third problem. Typically, the function of a trait is not just to produce some effect but to produce that effect *in the right situation*. The function of the stomach is to digest food, but only when there is non-toxic food available to it. The function of the white blood cells is to fight off infection, but only when there are infections to fight. Kingma (2010) calls this feature of functions their "situation specificity;" Neander (2012) makes a similar point when she talks about "response functions;" Garson and Piccinini (2014) talk about the "appropriate situation" for the performance of a function. So, can we formulate a general, context-independent method for identifying the appropriate situation for the performance of a function in any given case?

This problem matters because, like indeterminacy, it is crucial for deciding in any given case whether or not a trait token is dysfunctional. For example, my stomach is not digesting food right now. But it is not dysfunctional. Rather, I just have not eaten anything yet today. When we say that the stomach is dysfunctional, we are saying not only that it is not digesting food, but that, even if it were in an appropriate situation for digesting food, it *still* would not do so.

Kingma (2010) initially raised the problem of situation-specificity in a critique of Boorse's account of function. Her view was that Boorse's account of function—and statistical accounts of function quite generally—cannot make sense of the notion of dysfunction. She thought that, once we acknowledge the fact that an appropriate rate of function is relative to a situation, then we cannot distinguish (within the statistical account) functional from dysfunctional performances. Here is the argument, roughly. According the Boorse's theory (or at least a suitably-amended version of it) the function of a trait is its statistically typical contribution to fitness relative to a class of situations. So suppose that I have overdosed on paracetamol, but my stomach is functioning at a very low rate. Relative to that class of situations (that is, paracetamol overdose) my stomach is functioning at the rate that is typical for it in that situation. So we cannot say, contrary to normal medical judgment, that my stomach in that case is dysfunctional. Kingma's paper prompted a number of responses, including Hausman (2011), Kraemer (2013), Garson and Piccinini (2014), Boorse (2014), and Schwartz (2014).

In short, there is ample room for new theoretical innovations in the functions debate, as well as novel applications of the functions debate to ongoing scientific and philosophical problems. I hope this volume encourages philosophers and biologists to continue to advance that discussion.

References

Boorse, C. (2002). A rebuttal on functions. In A. Ariew, R. Cummins, & M. Perlman (Eds.), *Functions: New essays in the philosophy of psychology and biology* (pp. 63–112). Oxford: Oxford University Press.

Boorse, C. (2014). A second rebuttal on health. *Journal of Medicine and Philosophy, 39*, 683–724.

Buller, D. (1997). Individualism and evolutionary psychology (or, in defense of "narrow" functions). *Philosophy of Science, 64*, 74–95.

Dretske, F. (1986). Misrepresentation. In R. Bogdan (Ed.), *Belief: Form, content, and function* (pp. 17–36). Oxford: Clarendon Press.

Emiliani, V., Cohen, A. E., Deisseroth, K., & Häusser, M. (2015). All-optical interrogation of neural circuits. *Journal of Neuroscience, 35*, 13917–13926.

Garson, J. (2015). *The biological mind: A philosophical introduction*. London: Routledge.

Garson, J. (Submitted for publication). The developmental plasticity challenge to Wakefield's view. In L. Faucher & D. Forest (Eds.), *Defining mental disorder: Jerome Wakefield and his Critics*. Cambridge, MA: MIT Press.

Garson, J., & Piccinini, G. (2014). Functions must be performed at appropriate rates in appropriate situations. *British Journal for the Philosophy of Science, 65*, 1–20.

Goode, R., & Griffiths, P. E. (1995). The misuse of Sober's selection of/selection for distinction. *Biology and Philosophy, 10*, 99–108.

Hausman, D. (2011). Is an overdose of paracetamol bad for one's health? *British Journal for the Philosophy of Science, 62*, 657–668.

Hausman, D. (2012). Health, naturalism, and functional efficiency. *Philosophy of Science, 79*, 519–541.

Kingma, E. (2010). Paracetamol, poison, and polio: Why Boorse's account of function fails to distinguish health and disease. *British Journal for the Philosophy of Science, 61*, 241–264.

Kraemer, D. M. (2013). Statistical theories of functions and the problem of epidemic disease. *Biology and Philosophy, 28*, 423–438.

Millikan, R. G. (1984). *Language, thought, and other biological categories*. Cambridge, MA: MIT Press.

Neander, K. (1983). Abnormal Psychobiology. *Dissertation, La Trobe*.

Neander, K. (1995). Misrepresenting and malfunctioning. *Philosophical Studies, 79*, 109–141.

Neander, K. (2012). Teleosemantic theories of mental content. Stanford Encyclopedia of philosophy. http://plato.stanford.edu/entries/content-teleological/

Schwartz, P. H. (2007). Defining dysfunction: Natural selection, design, and drawing a line. *Philosophy of Science, 74*, 364–385.

Schwartz, P. H. (2014). Reframing the disease debate and defending the biostatistical theory. *Journal of Medicine and Philosophy, 39*, 572–589.

Sterelny, K. (1990). *The representational theory of mind: An introduction*. Oxford: Blackwell.